Applied Science Review™

Microbiology

 W9-BFF-706

Jay W. Wilborn, CLS, MEd
Director of Medical Laboratory Technology
Associate Degree Program
Garland County Community College
Hot Springs, Ark.

Springhouse Corporation
Springhouse, Pennsylvania

Staff

EXECUTIVE DIRECTOR, EDITORIAL
Stanley Loeb

DIRECTOR OF TRADE AND TEXTBOOKS
Minnie B. Rose, RN, BSN, MEd

ART DIRECTOR
John Hubbard

CLINICAL CONSULTANTS
Maryann Foley, RN, BSN; Cindy Tryniszewski, RN, MSN, CS

EDITORS
Diane Labus, David Moreau, Karen L. Zimmermann

COPY EDITORS
Melissa McGrath, Pamela Wingrod

DESIGNERS
Stephanie Peters (associate art director), Matie Patterson (senior designer)

COVER ILLUSTRATION
Scott Thorn Barrows

ILLUSTRATORS
Jacalyn Facciolo, Judy Newhouse, Stellarvisions

TYPOGRAPHY
David Kosten (director), Diane Paluba (manager), Elizabeth Bergman, Joyce Rossi Biletz, Phyllis Marron, Robin Mayer, Valerie L. Rosenberger

MANUFACTURING
Deborah Meiris (manager), Anna Brindisi, T.A. Landis

Library of Congress Cataloging-in-Publication Data
Wilborn, Jay W.
  Microbiology / Jay W. Wilborn.
    p. cm. — (Applied science review)
  Includes bibliographical references (p.   ) and index.
    1. Microbiology—Outlines, syllabi, etc.
  I. Title. II. Series.
QR62.W55 1993
576—dc20   92-17596
  ISBN 0-87434-457-3   CIP

Contents

Advisory Board and Reviewers vi
Acknowledgments and Dedication vii
Preface viii

1. Overview of Microbiology 1

2. Microscopy 6

3. Basic Techniques in Microbiology 14

4. Growth of Microorganisms 22

5. Identifying Microorganisms 31

6. Bacteria 36

7. Intermediate Organisms 51

8. Viruses 55

9. Fungi 64

10. Protozoa and Protozoa-like Organisms 72

11. Helminths 83

12. Infection and Epidemiology 93

13. Control of Microorganisms 98

Appendices

A: Glossary 106
B: Efficacy of Antibiotics Against Susceptible Pathogens 109

Selected References 116

Index 117

Advisory Board

Leonard V. Crowley, MD
Pathologist
Riverside Medical Center
Minneapolis;
Visiting Professor
College of St. Catherine, St. Mary's
Campus
Minneapolis;
Adjunct Professor
Lakewood Community College
White Bear Lake, Minn.;
Clinical Assistant Professor of Laboratory
Medicine and Pathology
University of Minnesota Medical School
Minneapolis

David W. Garrison, PhD
Associate Professor of Physical Therapy
College of Allied Health
University of Oklahoma Health Sciences
Center
Oklahoma City

Charlotte A. Johnston, PhD, RRA
Chairman, Department of Health
Information Management
School of Allied Health Sciences
Medical College of Georgia
Augusta

Mary Jean Rutherford, MEd, MT(ASCP)SC
Program Director
Medical Technology and Medical
Technicians—AS Programs;
Assistant Professor in Medical Technology
Arkansas State University
College of Nursing and Health
Professions
State University

Jay W. Wilborn, CLS, MEd
Director, MLT-AD Program
Garland County Community College
Hot Springs, Ark.

Kenneth Zwolski, RN, MS, MA, EdD
Associate Professor
College of New Rochelle
School of Nursing
New Rochelle, N.Y.

Reviewers

Ellen P. Digan, MT(ASCP)MA
Professor and Coordinator
Medical Laboratory Technology Program
Manchester (Conn.) Community College

Carol McLimans, BS, MT(ASCP)SM
Teaching Technologist in Clinical
Microbiology
Mayo Clinic
Rochester, Minn.

Microbiology

Acknowledgments and Dedication

Thanks to Ruth Wallace for help in preparation of the manuscript. And my utmost appreciation to the editors, reviewers, and staff of Springhouse Corporation for making this publication possible.

To my favorite microbiologist, Ellen, and to all the microbiology students who, through the years, have made the effort worthwhile.

Preface

This book is one in a series designed to help students learn and study scientific concepts and essential information covered in core science subjects. Each book offers a comprehensive overview of a scientific subject as taught at the college or university level and features numerous illustrations and charts to enhance learning and studying. Each chapter includes a list of objectives, a detailed outline covering a course topic, and assorted study activities. A glossary appears at the end of each book; terms that appear in the glossary are highlighted throughout the book in boldface italic type.

Microbiology provides conceptual and factual information on the various topics covered in most microbiology courses and textbooks and focuses on helping students to understand:

• the nature and significance of microbiology as a field of study
• the use of microscopes and culture and staining techniques to study and describe specimens
• the various processes — metabolism, food production, respiration, reproduction, and enzyme function — that play a role in microbial growth
• the methods that scientists use to name and classify microorganisms
• the identification, location, importance, and taxonomic classifications of selected bacteria, intermediate organisms, viruses, fungi, protozoa, and helminths
• the development and transmission of infectious diseases and the control measures used to inhibit or destroy pathogenic microorganisms.

1

Overview of Microbiology

Objectives

After studying this chapter, the reader should be able to:
• Define microbiology, pathogen, and symbiosis.
• List and describe the six types of microorganisms.
• Summarize the five basic methods used to study microorganisms.
• Describe the three types of symbiotic relationships.
• Name the two major subdivisions of microbiology.
• Discuss the historical evolution of microbiology.

I. Microorganisms

A. General information

    1. ***Microorganisms,*** also called microbes or germs, are living organisms that are too small to be seen without the aid of a ***microscope***

    2. Microorganisms have subcellular, single-cell, or multicell forms

       a. Subcellular forms include viruses, which exist in other living ***cells***

       b. Single-cell forms include bacteria and protozoa, which carry on all the processes of life as independent organisms

       c. Multicell forms include some fungi and helminths, which have stages of growth large enough to be seen without the aid of a microscope

    3. Microorganisms live in soil, air, fresh and salt water, other microorganisms, plants, animals, and humans

    4. Microorganisms are involved in such processes as bread leavening, beer and wine making, food preservation and spoilage, and the transmission of infections and contagious diseases

B. Types of microorganisms

    1. Many types of microorganisms exist, including bacteria, intermediate organisms, viruses, yeasts and molds, protozoa, and ***helminths***

    2. *Bacteria* are a large and diverse group of microorganisms about 1 micrometer in diameter

       a. Their characteristics are similar to those of single-cell plants in that they have cell walls and may produce their own food

       b. Unlike single-cell plants, bacteria lack a definite cell ***nucleus*** and membrane-bound ***organelles***

3. *Intermediate organisms* have characteristics that are simpler than those of bacteria yet more complex than those of viruses
   a. These organisms may lack cell walls
   b. They frequently depend on other organisms for nourishment
4. *Viruses,* the smallest microorganisms, have the simplest composition and life cycles
   a. Viruses exist as complex molecules
   b. They cannot carry on any life processes independently and must depend on other organisms
5. *Yeasts and molds,* which are classified as fungi, include both single-cell and multicell organisms
6. *Protozoa* and related organisms are animals and plants that have definite nuclei
7. *Helminths* are wormlike, multicell organisms
   a. Helminths have microscopic stages as ova and larvae but may have adult stages large enough to be seen without the aid of a microscope
   b. They are highly organized, with specialized tissues and organ systems to carry on complex life cycles, including free-living and parasitic forms

C. Symbiosis

1. Microorganisms exist in close relationships with other microorganisms or with a host (any organism that provides a site for the growth of another organism)
2. The three types of **symbiosis** are commensalism, mutualism, and parasitism
   a. *Commensalism* is a relationship in which one organism receives its nourishment from another; the host organism does not benefit from nor is it harmed by this relationship (for example, bacteria on the hair and skin do not harm their hosts)
   b. *Mutualism* is a relationship in which two organisms benefit each other by supplying nourishment or providing protection (for instance, bacteria living in the human intestinal tract aid in the metabolism of vitamins)
   c. *Parasitism* is a relationship in which one organism benefits while the other is harmed (for example, tapeworms deprive their host of needed food)
      (1) When the harm results in disease, the **parasite** is called a **pathogen** (the bacterium *Streptococcus pyogenes,* a pathogen, causes strep throat)
      (2) Organisms that are pathogenic only if conditions in the host are right (opportune) are called **opportunists** (vaginal infections can be caused by the yeast *Candida albicans)*

D. Studying microorganisms

1. To study microorganisms, microbiologists use microscopes, culture techniques, identification procedures, and measures for limiting or preventing growth
   a. Microscopes are instruments that produce enlarged images of objects too small to be seen with the unaided eye
   b. Culture techniques provide the nutrients and atmospheric conditions necessary to grow microorganisms for study
   c. Identification procedures include recognition of the size, shape, organization, chemical composition, and antigenic makeup of microorganisms

        d. Measures for limiting or preventing growth of microorganisms allow for controlled study and treatment of disease

  2. ***Microbiology*** is subdivided into general microbiology and applied microbiology

      a. *General microbiologists* use basic biologic techniques to study microorganisms (see Chapters 2 through 5)

         (1) General microbiologists study representative microorganisms of each type (see Chapters 6 through 11)

         (2) They seek a basic understanding of microorganisms as a foundation for more advanced study

      b. *Applied microbiologists* use the principles of microbiology in medical, industrial, agricultural, and environmental applications

         (1) Medical microbiologists study the causes, transmission, and treatment of contagious diseases (see Chapters 12 and 13)

         (2) Industrial microbiologists are involved in processes ranging from the manufacture of paper and clothing to the production of chemicals

         (3) Agricultural microbiologists are concerned with plant nutrition and the making and preserving of foods and beverages

         (4) Environmental microbiologists apply microbiologic techniques to sanitation and sewage treatment, pest control, and cleanup of oil spills

  3. Specialized areas of microbiology include ***bacteriology*** (the study of bacteria), ***mycology*** (the study of yeasts and molds), ***virology*** (the study of viruses), and ***parasitology*** (the study of parasites)

II. Significant Historical Events in Microbiology

A. General information

  1. Scientists have studied microorganisms for more than 400 years

  2. Their study has been enhanced by the invention of such instruments as the microscope

  3. From the 16th century to the present, many theories have been developed about the growth and control of microorganisms

B. 16th century

  1. In 1546, Girolamo Fracastoro proposed the theory of contagious diseases

      a. He believed that diseases were spread through contact between individuals

      b. He developed this theory while treating cases of syphilis

  2. In 1590, Johannes and Zacharias Janssen invented the first compound microscope (one having two sets of lenses)

      a. The Janssens used sunlight to illuminate the object under study

      b. Their microscope achieved magnifications of 10 to 100 times the object's actual size

C. 17th century

  1. In 1665, Robert Hooke advanced the cell theory of biology

      a. Hooke studied fungi and discovered cells

      b. He proposed that cells are the basic unit of all living things

    2. In 1674, Antonie van Leeuwenhoek used a microscope to describe
microorganisms
       a. Leeuwenhoek developed lenses capable of magnifying an object up to 270
times its actual size
       b. He examined pond water and mouth scrapings to discover and record
microorganisms

D. 18th century
    1. From 1711 to 1776, Louis Jobbt, John Needham, and Lazzaro Spallanzani
performed experiments involving spontaneous generation (life developing
from nonliving materials)
       a. In these experiments, various media grew microorganisms only when
exposed to air, a source of bacteria and molds
       b. They theorized that all life came from existing life forms
    2. In 1798, Edward Jenner developed a *vaccine* for smallpox
       a. Jenner noted the similarity between smallpox and cowpox
       b. He discovered that *vaccination* with the live virus of cowpox protected
individuals from smallpox

E. 19th century
    1. In 1866, Louis Pasteur developed the process now known as pasteurization
       a. Pasteurization uses moderate heat (below boiling) to destroy harmful
microorganisms
       b. This process does not chemically alter the substance being pasteurized
    2. In 1867, Joseph Lister began using *antiseptics* in surgery
       a. Lister experimented with the then-developing concept that microorganisms
cause disease
       b. He used phenol as an antiseptic on surgical wounds
    3. Between 1880 and 1885, Louis Pasteur developed vaccines
       a. Pasteur worked with cholera and anthrax (bacterial infections) and rabies
(a viral infection)
       b. He found that attenuated (weakened) organisms could not cause infection
but would produce immunity
    4. In 1882, Robert Koch discovered the cause of tuberculosis
       a. Tuberculosis is caused by a bacterium
       b. Koch's studies led to the germ theory of disease
    5. In 1900, Walter Reed proved that mosquitoes transmitted yellow fever
       a. Yellow fever is caused by a virus
       b. The infection caused many deaths in the United States and Central
America before being controlled through eradication of the mosquitoes
that carried it

F. 20th century
    1. In 1908, Paul Ehrlich introduced chemotherapy
       a. Ehrlich sought a treatment for syphilis
       b. He developed Salvarsan, an arsenic compound that was specific and
effective

2. In 1929, Alexander Fleming discovered **antibiotics**
    a. He made this discovery while working with a culture of bacteria that had become contaminated with the mold *Penicillium*
    b. After noting that the mold inhibited the growth of the bacteria, Fleming isolated the antibiotic penicillin
3. In 1933, Ernst Ruska developed the electron microscope
    a. To illuminate objects for study, Ruska used a beam of electrons rather than sunlight, focusing the beam with electrical lenses
    b. The electron microscope has permitted scientists to magnify an object millions of times its actual size
4. In 1952, Jonas Salk tested a vaccine against polio
    a. Polio is caused by a virus
    b. Salk used dead viruses in his vaccine to provide immunity against the disease
5. From 1953 to the present, researchers have continued to apply the techniques of microbiology in order to define new infections—such as those that cause Legionnaire's disease and acquired immunodeficiency syndrome—and to pursue effective treatments and methods of prevention

Study Activities

1. Write an essay that addresses the role of microorganisms in food preservation and food spoilage.
2. Describe at least five ways in which microorganisms benefit humans. How many benefits did you know about before studying the issue?
3. Identify two organisms that share a symbiotic relationship of mutualism, and summarize the various benefits that each organism derives from the other.
4. Interview an environmental microbiologist about the techniques now being used to clean up oil spills. How effective are these techniques compared with those used 10 years ago? What other environmental problems are microbiologists currently investigating? Are any definitive solutions to these problems likely to emerge in the next 10 years?
5. Form a discussion group with your classmates to research and debate the following questions: How did Louis Pasteur influence the work of Joseph Lister and Robert Koch? Which aspects of contemporary life would be different without the efforts of Walter Reed, Paul Ehrlich, and Alexander Fleming?
6. List four significant discoveries in microbiology that would not have been possible before Ernst Ruska developed the electron microscope in 1933.

2

Microscopy

Objectives

After studying this chapter, the reader should be able to:
• Define microscopy.
• List and describe the five types of microscopes.
• Identify the parts of a compound light microscope.
• Discuss the preparation of wet and dry smears.
• Describe the procedure for using a light microscope.

I. Types of Microscopes

A. General information
1. Microscopy is the skilled use of microscopes — instruments designed to obtain an enlarged image of a small object or specimen
2. Microscopes allow the microbiologist to study microorganisms and to gather information about their characteristics
3. The first microscopes, developed in the 16th century, used simple *lenses* to focus ordinary light
   a. At first, magnification was limited to about ten times (10×) the object's actual size
   b. Improvements eventually permitted magnifications of 270× to 400×
4. Advances in technology permitted the use of electromagnetic radiation other than light to produce an enlarged image of a specimen
5. Cost, availability, ease of use, and desired magnification are the primary considerations when selecting a particular type of microscope
6. Modern microscopes include the light microscope, the ultraviolet microscope, the fluorescence microscope, the electron microscope, and the acoustic microscope

B. Light microscope
1. A light microscope uses ordinary *white light* to view microorganisms
   a. The light can be passed directly through the specimen or around the edge of the specimen *(darkfield microscopy)*
   b. Polarized light, produced by passing ordinary light through two filters, can be used to provide greater detail *(phase-contrast microscopy)*
2. A light microscope allows the microscopist to view an enlarged image directly with the eyes

   3. This type of microscope can magnify a specimen 1,000 times its actual size
   4. Light microscopes use one or more lenses to focus the light
      a. Simple light microscopes use one lens
      b. Compound light microscopes use two sets of lenses

C. Ultraviolet microscope
   1. An ultraviolet microscope uses shorter **_wavelength ultraviolet (UV) light,_** rather than white light, to view microorganisms
   2. An ultraviolet microscope can be used to resolve objects smaller than those visible with a light microscope
   3. The image produced by this type of microscope is recorded on photographic film; the microscopist does not view the image directly
   4. The magnification possible with an ultraviolet microscope is approximately the same as that of a light microscope

D. Fluorescence microscope
   1. A **_fluorescence_** microscope also uses ultraviolet light
      a. The specimen absorbs light produced by the fluorescence microscope
      b. The absorbed light causes the specimen to fluoresce (give off light), thereby making it visible to the microscopist
   2. Use of this type of microscope commonly involves the application of fluorescent dyes to stain the specimen
      a. Staining facilitates the detection and identification of a particular cell type
      b. The procedure can be learned quickly, even by a novice microscopist
   3. A fluorescence microscope allows the microscopist to view the specimen directly
   4. It can magnify a specimen up to 1,000 times its actual size

E. Electron microscope
   1. An electron microscope uses a focused beam of electrons, rather than light, to view the specimen
      a. Electrons pass through the specimen and are focused electrically
      b. The specimen may be a large, complex molecular form, such as a virus; a single cell; an organelle; tissue; or a worm, insect, or similar object of interest
      c. Preparation of specimens commonly involves freezing and treatment with salts of heavy metals, such as gold and uranium, to produce shadows on the images
   2. The microscopist views the image on a television screen or photographic film rather than viewing the specimen directly
   3. An electron microscope can magnify a specimen millions of times its actual size
   4. Scanning electron microscopes provide a three-dimensional (and, hence, more realistic) image of a specimen

F. Acoustic microscope
   1. An acoustic microscope uses computer-analyzed sound waves to view the specimen

2. As with the electron microscope, an acoustic microscope generates an image electronically on a television screen
3. It can magnify specimens up to 5,000 times their actual size

II. Parts of a Compound Light Microscope

A. General information
1. A compound light microscope is the most commonly used instrument in microbiology
2. The lenses of this microscope are arranged to bring an enlarged image to the eyes of the microscopist (see *Parts of a Compound Light Microscope*)

B. Lenses
1. Lenses are pieces of transparent material, such as glass, that are shaped to form an image by focusing rays of light
2. Lenses are critical to the magnifying power and the resolving power of a microscope
   a. ***Magnifying power*** is the ratio of apparent image size (the size of the object as seen through the microscope) to actual size (without magnification)
      (1) To determine the effective magnification of a specimen, the microscopist multiplies the magnifying power of the ocular (usually 10×) by the magnifying power of the objectives (usually 5×, 10×, 45×, or 100×)
      (2) This results in an effective magnification of 50× to 1,000× actual size
   b. ***Resolving power*** is the ability to distinguish the smallest of details
      (1) It allows the microscopist to determine, for example, whether two points are separate or part of a whole
      (2) The unaided human eye has a resolving power of about 100 micrometers
      (3) Resolving power is a function of the wavelength of the energy used to illuminate the specimen
         (a) A light microscope can resolve objects as small as 0.2 micrometer
         (b) An ultraviolet microscope can resolve objects as small as 0.1 micrometer
         (c) An electron microscope can resolve objects as small as 0.001 micrometer

C. Oculars
1. Each ocular, or eyepiece, contains one set of lenses that bring focused light to the eye
2. A monocular microscope has one ocular; a binocular microscope has two
3. The lenses of most oculars magnify an object 10 times its actual size

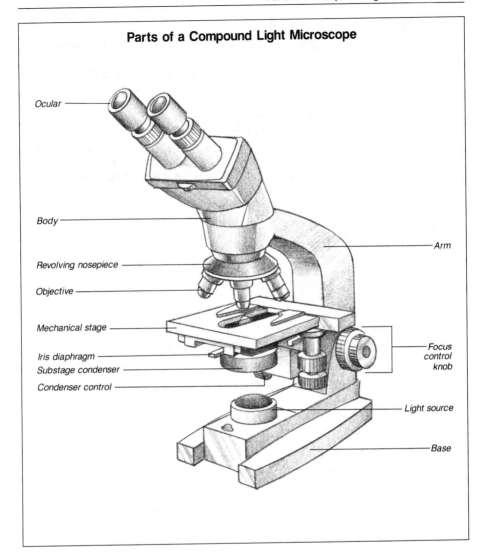

Parts of a Compound Light Microscope

Ocular

Body

Arm

Revolving nosepiece

Objective

Mechanical stage

Focus control knob

Iris diaphragm

Substage condenser

Condenser control

Light source

Base

D. Body
    1. The body of the light microscope supports the lens system
    2. It also provides a path from the light source to the eyes

E. Arm
    1. The arm supports the body
    2. It is used for carrying the microscope

F. Revolving nosepiece
    1. The revolving nosepiece holds the objectives
    2. It allows the microscopist to choose the level of magnification

G. Objective

1. The objective of a light microscope contains the primary lens that magnifies the specimen
2. It collects light from the specimen to produce magnifications of $5\times$, $10\times$, $45\times$, or $100\times$ the object's actual size, depending on which objective is selected
3. Each objective is labeled with its magnification and type
   a. One type of objective, the **parfocal,** does not require refocusing when the microscopist switches from a low-powered lens to a high-powered one
   b. An oil-immersion objective, used to view a specimen that has been immersed in optical-quality oil, bridges the space between the specimen and the lens, thereby reducing the amount of scattered light and improving resolving power

H. Mechanical stage

1. The mechanical stage is the platform that supports the specimen and the microscope slide
2. This part of the light microscope allows the microscopist to move the specimen within the field of view

I. Substage condenser

1. The substage condenser is the lens located beneath the mechanical stage
2. It condenses light before it passes through the specimen

J. Iris diaphragm

1. The iris diaphragm adjusts the amount of light passing through the substage condenser
2. The iris opens and closes like the shutter or diaphragm of a camera

K. Condenser control

1. The condenser control raises and lowers the substage condenser
2. It allows the microscopist to vary the intensity of light focused on the specimen

L. Focus control knob

1. Used for coarse and fine adjustment, the focus control knob is found on the arm of the microscope
2. Turning the knob alters the distance between the objective and the specimen, enabling the microscopist to achieve a sharp, clear image

M. Light source

1. The light source emanates from the base of the microscope, beneath the substage condenser
2. It provides bright, white light for viewing the specimen

N. Base

1. The base supports the microscope
2. Along with the arm, it is used to carry the microscope

III. Preparing Specimens for the Light Microscope

A. General information

1. Microorganisms can be viewed with a light microscope by placing them on a microscope slide
   a. Slides are made of thin, clear glass or plastic
   b. The specimen may be topped with a cover slip, a thin piece of glass
2. *Specimens* can be placed on slides as wet or dry *smears* (small amounts of the specimen)
3. Microorganisms can be preliminarily identified by combining the results of wet smears, dry smears, and descriptions of the organism's *morphology* (shape and structure) as detected by microscopy

B. Wet smears

1. Wet smears, or preparations, are used to view living microorganisms quickly, before changes (such as from handling or drying) can occur
2. The microscopist prepares a wet smear by placing a small drop of liquid, which contains the microorganism, onto a slide
   a. Liquids used include water, saline solution, urine, and other body fluids
   b. Large numbers of bacteria in body fluids usually indicate infection
3. Microorganisms in wet smears are in a nearly natural living state
   a. For this reason, the organism's morphology can be described
   b. The organism's *motility* (ability to move) also can be determined

C. Dry smears

1. Dry smears are used when dead organisms can be observed with stains at higher magnifications
2. The microscopist prepares a dry smear by placing a specimen on a microscope slide and allowing it to dry
3. Dry smears of specimens are stained with one or more stains
   a. The type of stain used depends on the information sought by the microscopist
      (1) The Gram stain separates common bacteria and yeast into two main categories
      (2) The capsule stain detects and identifies bacterial capsules
      (3) The acid-fast stain detects chemical differences in the cell walls of different organisms
   b. Stains allow the microscopist to better visualize the microorganism's characteristics, such as size, shape, organization, color, and presence of specialized structures
   c. As an example, bacteria from the lungs that stain red with the acid-fast stain are most likely the cause of tuberculosis
4. Drying and staining affect the microorganisms being studied
   a. The dry smear process kills the microorganisms in the specimen
   b. Other changes may occur, including changes in the organism's size and shape and the ability to use specific stains

IV. Using a Light Microscope

A. General information

1. The microscope must be kept clean and care should be taken to prevent damage to its delicate parts
2. Before viewing a specimen, the microscopist should place the instrument on a stable surface at a comfortable height

B. Procedure for using a light microscope

1. The microscopist turns on the light source and then places a wet preparation or stained dry smear on the mechanical stage, being careful not to hit the slide against the objectives
2. While grasping the revolving nosepiece, the microscopist rotates a low-power objective in place over the specimen
   a. By looking through the ocular, the microscopist can focus the specimen, first with the coarse adjustment and then with the fine adjustment
   b. Low light levels should be used with low-power objectives or when examining wet preparations; bright light tends to wash out the image, hindering visualization
3. After viewing the specimen with the low-power objective, the microscopist may rotate a high-power objective into place
   a. Changing the objectives may require the microscopist to readjust the focus; this is unnecessary with a parfocal microscope, because the specimen remains in focus when the objectives are rotated
   b. A high-power objective may require a brighter light for optimal viewing of the specimen
4. The microscopist notes the details of the specimen and records all observations
5. After rotating the objectives out of the way, the microscopist can remove the specimen and discard it or store it safely
6. Finally, the microscopist should wipe the objectives and oculars clean with lint-free lens paper and turn off the light source

C. Procedure for increasing magnification

1. A microscopist who is viewing stained dry smears can use oil immersion techniques to increase the microscope's magnification power
2. While the slide is in place on the mechanical stage, the microscopist rotates the objectives out of the way and places a small drop of optical grade immersion oil on the slide
3. The highest power objective is then rotated into place and lowered until it contacts the immersion oil
   a. Contact between the objective and the immersion oil creates a continuous path of light
   b. This continuous path minimizes the loss of light caused by scattered light rays, thereby achieving optimum resolving power and magnification

Study Activities

1. Research the advantages and limitations of darkfield microscopy and phase-contrast microscopy. Then determine which type of microscope would permit better visualization of an unstained live specimen. Defend your response.

2. Draw a rough sketch of a compound light microscope. Label all the parts and describe their functions.

3. Calculate the effective magnification produced by a compound light microscope if the magnifying power of the ocular lens is $10\times$ and the magnifying power of the objective lens is $45\times$. How much of an increase in magnification would you achieve by changing the objective lens to $100\times$?

3

Basic Techniques in Microbiology

Objectives

After studying this chapter, the reader should be able to:
• Define aerobe, anaerobe, inoculation, and medium.
• Explain the difference between a pure culture and a mixed culture.
• Describe the basic structure of a bacterial cell.
• Name six types of culture media.
• Identify five types of special stains and describe their uses.

I. Introduction

A. General information

1. Microbiologists use microscopes, cultures, and stains to study microorganisms
    a. Microscopes are used to observe microorganisms directly
    b. Cultures are used to measure the characteristics common to all living organisms, including reproduction, ability to pass traits from one generation to the next, and organization into structural forms, from simple to complex
    c. Stains, either simple dyes or a mixture of dyes, are used to obtain a wide array of useful morphologic information
2. To ensure the growth and survival of microorganisms, microbiologists must provide proper atmospheric conditions
3. The techniques of microscopy, culturing, staining, and meeting atmospheric conditions can be illustrated by their use in the study of bacteria

B. Bacterial shapes and organization

1. Bacteria are shaped as spheres (cocci), rods (bacilli), or curved forms and spirals (spirochetes), as shown in *Bacterial Shapes*
2. Bacteria are organized in various ways
    a. They can occur singly when they separate after reproducing
    b. They can occur in pairs (indicated by the prefix *diplo-*)
    c. They can appear as chains (indicated by the prefix *strepto-*)
    d. They can appear as clusters or irregular masses (indicated by the prefix *staphylo-*)
    e. They can be found in regular packets of four, eight, or sixteen organisms

Bacterial Shapes

As shown in these illustrations, bacteria have three basic shapes: rods (bacilli), spheres (cocci), and spirals.

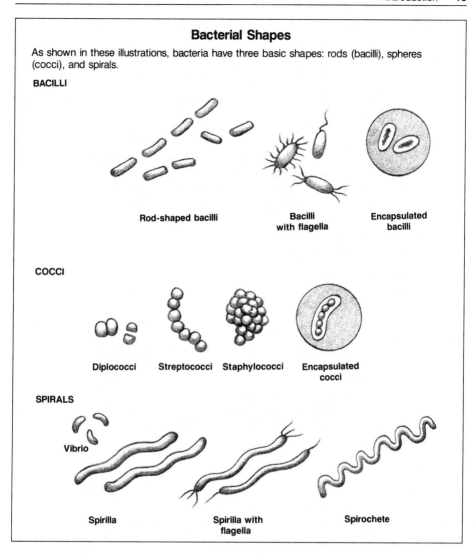

BACILLI

Rod-shaped bacilli   Bacilli with flagella   Encapsulated bacilli

COCCI

Diplococci   Streptococci   Staphylococci   Encapsulated cocci

SPIRALS

Vibrio

Spirilla   Spirilla with flagella   Spirochete

C. Bacterial cell structure

1. A bacterial cell is protected by a rigid *cell wall* composed of polysaccharides, lipids, lipoproteins, peptidoglycans, and organic acids
2. To protect the cell from other microorganisms or the host's defenses, the cell wall may be covered by a *slime layer* of mucoid material that diffuses from the cell or by a thick, mucoid *capsule;* some cells are protected by both a slime layer and a mucoid capsule
3. **Pili** are fingerlike projections used by the cell to anchor itself to surfaces in its environment
4. Through their whipping and twisting actions, **flagella** permit the bacterium to move or swim

5. The interior cell wall is lined by a cell membrane, a semipermeable layer that insulates the cell from environmental changes, such as variations in salt and water content of the surroundings
6. The inside of the cell, the cytoplasm, is composed of water, amino acids, nucleic acids, protein, carbohydrates, lipids, and electrolytes
7. Embedded in the cytoplasm are functional units called *organelles,* which play a role in metabolism, reproduction, and growth
8. The cell's characteristics are determined by genes, which are located on *chromosomes*
9. Chromosomes are contained in a saclike nucleus in *eukaryotic* (nucleated) cells but float free within the cytoplasm of *prokaryotic* (nonnucleated) cells
   a. Bacteria and blue-green algae are prokaryotic organisms
   b. Fungi, protozoa, plants, and animals have eukaryotic cells

II. Culturing Microorganisms

A. General information
1. Bacteria and other microorganisms can be studied in the microbiology laboratory by growing, or culturing, them
2. Culturing microorganisms in a *medium* (nutritive substance) is relatively easy and inexpensive
   a. A pure culture contains only one species, or type, of microorganism
   b. A mixed culture contains more than one species
3. Microorganisms also can be grown in living cells and other organisms or *hosts*
   a. Cell cultures and animal host cultures require more time and technical expertise and are more expensive than cultures in media
   b. These types of cultures are used for viruses and intermediate organisms that are parasitic in other cells

B. Reproduction of microorganisms
1. Microorganisms reproduce when they are situated in a suitable medium or host
   a. They can reproduce by *fission* (splitting), which produces two equal daughter cells (binary fission), as in Streptococci
   b. They can reproduce by budding, a method in which a daughter cell breaks off from a parent cell, as in Candida
   c. In cell cultures or animal host cultures, death or disease signals bacterial reproduction
2. Some bacteria reproduce as often as once every 20 minutes; other bacteria may require as long as 6 hours
3. Within 18 hours to 6 weeks, depending on the type of microorganism, cultures of bacteria on media can produce visible colonies composed of millions of cells

C. Culture media
1. Microorganisms cannot be cultured unless certain conditions are met
   a. The medium must contain water, which is essential to all living things
   b. The microorganism must be provided with nutrients, such as minerals, vitamins, and sources of carbon, nitrogen, and other elements

    c. A proper atmosphere must be maintained
      (1) The humidity must be kept between 40% and 70% to prevent the
          microorganism from drying out
      (2) A proper **pH** level (neutral to alkaline for bacteria; acidic for fungi) is
          crucial
      (3) Appropriate temperatures are necessary
          (a) Cryophiles, or psychrophiles, grow in the cold (from $-15°$ to
               $2°$ C); this group includes Listeria
          (b) Mesophiles grow at moderate temperatures (from $20°$ to $40°$
               C); this group includes Staphylococcus
          (c) Thermophiles grow at high temperatures (from $45°$ to $100°$ C);
               this group includes *Bacillus stearothermophilus*
    d. An adequate amount of time (4 to 18 hours for the first signs of the
      fastest-growing organisms) is needed
2. Media can be liquid, semisolid, or solid to meet the physical requirements of the
   microorganisms or the culture requirements of the microbiologist
3. Media are either complex or chemically defined
    a. Complex media contain natural substances, such as blood or eggs, that
      are not thoroughly analyzed
    b. Chemically defined media have all ingredients and their amounts specified
4. Several different types of culture media are used (see *Common Culture Media,*
   page 18)
    a. *Nutrient media,* which include nutrient broth (a liquid culture medium) and
      nutrient agar (a solid culture medium), provide only basic requirements
      for growth
    b. *Enriched media* use nutrient media as a base, along with such additives
      as blood, serum, meat, or eggs to enhance growth
    c. *Selective media* contain such additives as salt, dyes, and antibiotics,
      which allow certain organisms to grow while inhibiting others; eosin-
      methylene blue (EMB) agar, colistin-nalidixic acid (CNA) agar, and 6.5%
      sodium chloride broth are examples of selective media
    d. *Differential media* yield identifying characteristics from physical or
      chemical reactions of the organisms, such as the degree of hemolysis on
      blood agar or the production of acids, alcohols, or gases
    e. *Enrichment media* promote the growth of certain organisms while limiting
      the growth of others; gram-negative broth is an example of an
      enrichment medium
    f. *Transport media* keep microorganisms alive, without allowing them to
      reproduce, while they are transported from the collection site to the
      microbiology laboratory; Amies medium is a transport medium

D. Obtaining a culture

1. To produce a culture, a microbiologist first selects an appropriate medium
    a. A solid medium, such as blood **agar,** is placed on a plate, or Petri dish
    b. A liquid or semisolid medium, such as gram-negative **broth,** is placed in a
      tube

Common Culture Media

The chart below lists common culture media and their primary uses, including those for solid (S), liquid (L), complex (C), chemically defined (CD), differential (D), enrichment (Em), enriched (En), and selective (Se) media.

NAME	TYPE	USE
Blood agar	S, C, En, D	All-purpose medium for common bacteria and some fungi
Brain-heart infusion (BHI) agar	S, C, En	All-purpose medium for common bacteria
Brain-heart infusion (BHI) broth	L, C, En	All-purpose medium for common bacteria
Chocolate agar	S, C, En	All-purpose medium for common bacteria
Colistin-nalidixic acid (CNA) agar	S, C, En, Se, D	Selective medium for gram-positive cocci
Eosin–methylene blue (EMB) agar	S, CD, Se, D	Selective medium for gram-negative bacilli
Gram-negative (GN) broth	L, CD, Se, Em	Selective medium for pathogenic gram-negative bacilli
Hektoen enteric (HE) agar	S, CD, Se, D	Selective medium for gram-negative bacilli
Kligler iron agar	S, CD, D	Differential medium for gram-negative bacilli
Löwenstein-Jensen culture	S, C, Se, D	Selective medium for bacilli of tuberculosis and related bacteria
MacConkey (MC) agar	S, CD, Se, D	Selective medium for gram-negative bacilli
Mannitol salt agar	S, CD, Se, D	Selective medium for staphylococci
Middlebrook-Cohn agar	S, CD, Se, D	Selective medium for bacilli of tuberculosis and related bacteria
Mycosel agar	S, CD	All-purpose medium for fungi
Phenylethyl alcohol (PEA) agar	S, CD, Se	Selective medium for gram-positive cocci
Sabouraud dextrose agar	S, C	All-purpose medium for fungi
Thayer-Martin (TM) agar	S, C, En, Se	Selective medium for gram-negative cocci of gonorrhea and related bacteria
Thioglycollate broth	L, CD, D	All-purpose medium for growth of anaerobic, microaerophilic, facultative, and aerobic bacteria
Xylose-lysine-deoxycholate (XLD) agar	S, CD, Se, D	Selective medium for gram-negative bacilli

      c. The medium is allowed to reach room temperature if it had been refrigerated

  2. The microbiologist then introduces microorganisms into or onto the medium to obtain a culture; this process is called *inoculation*

      a. A sterile wire needle or loop is used to place microorganisms onto the medium

      b. For media in Petri dishes, the **inoculum** is spread out, or streaked, over the medium in three stages; at some point along the streak pattern, isolated microorganisms will grow in pure culture

III. Techniques for Achieving Proper Atmospheric Conditions

A. General information

  1. Microorganisms differ in their need for atmospheric gases (see *Atmospheric Requirements and Sources of Common Bacteria*)

  2. Oxygen is essential to some microorganisms but not to others

      a. **Aerobes** and *microaerophiles* require oxygen in their atmosphere

      b. **Anaerobes** do not require oxygen

  3. Increased carbon dioxide (5% to 10%) is necessary for the growth of organisms referred to as *capnophiles*

  4. **Obligate** organisms have an absolute need for a particular atmosphere; for example, an obligate aerobe is an organism that requires oxygen

  5. **Facultative** organisms can grow under various conditions; for example, a facultative anaerobe is an aerobe that can grow anaerobically

B. Procedure for maintaining atmospheric conditions

  1. To maintain *aerobic* conditions, the microbiologist uses an atmosphere composed of room air, 21% oxygen, and less than 5% carbon dioxide

---

Atmospheric Requirements and Sources of Common Bacteria

The chart below lists common bacteria, their requirements for atmospheric conditions, and the sources from which they are usually isolated.

BACTERIA	REQUIREMENTS	SOURCES
Pseudomonas	Aerobic, obligate	Soil, water, human and animal infections
Clostridium, Bacteroides, Pepto-coccus	Anaerobic, obligate	Soil, human and animal gastro-intestinal tracts, wounds (puncture wounds, trauma, bites)
Escherichia coli, staphylococci, streptococci	Facultative	Humans and animals
Campylobacter, Helicobacter, Lactobacillus	Microaerophilic	Human and animal gastrointestinal tracts
Haemophilus, Neisseria gonor-rhoeae, mycobacteria	Capnophilic	Meningitis (*Haemophilus*), gonorrhea (*Neisseria gonorrhoeae*), and tuberculosis (mycobacteria)

2. To maintain *microaerophilic* conditions, the microbiologist partially displaces room air with nitrogen, hydrogen, or carbon dioxide; creates a chemical reaction between oxygen and hydrogen or carbon; uses reducing agents (chemicals, such as thioglycollate broth) to react with oxygen; and uses less than 10% oxygen

3. To maintain *anaerobic* conditions, the microbiologist completely replaces room air with nitrogen, hydrogen, or an inert gas; creates a chemical reaction between oxygen and hydrogen or carbon; uses reducing agents to react with oxygen; and ensures that no free oxygen remains in the atmosphere

4. To maintain *capnophilic* conditions, the microbiologist uses a candle jar (a sealed container in which a lighted candle will combust oxygen and carbon to carbon dioxide) and maintains carbon dioxide levels of 5% to 10%

IV. Staining Techniques

A. General information

1. A stain is a substance used to color or define cells or cell structures for ease of microscopic observation
2. Stains can be positive or negative
   a. A *positive* stain is adsorbed directly by the object under study
   b. A *negative* stain, also called a relief stain, outlines a structure, such as a capsule, that cannot be directly stained
3. Stains are classified as simple or special, depending on the number of steps or components in the process

B. Simple positive stains

1. To prepare a simple positive stain, the microbiologist first places a smear on a slide and allows it to dry
2. The dry smear is then flooded with a stain, such as crystal violet, carbolfuchsin, methylene blue, or safranin
3. The excess stain is washed off with water
4. After it dries, the stained smear is examined under the microscope

C. Special stains

1. A *Gram stain* stains many bacteria either blue (gram-positive) or red (gram-negative)
   a. To prepare a Gram stain, the microbiologist covers a smear with crystal violet stain and iodine, which stains all bacteria blue
   b. The microbiologist then decolorizes the smear with acetone-alcohol; during this process, some bacteria lose their blue color, depending on the chemical makeup of their cell walls
   c. Next, the microbiologist applies the red stain safranin to stain the decolorized bacteria
   d. The Gram stain reaction, along with bacterial morphology, provides the microbiologist with clues to the bacteria's identification

2. An *acid-fast stain* stains bacteria, especially those that cause tuberculosis, either red (acid-fast positive) or green to blue (acid-fast negative)
   a. To prepare an acid-fast stain, the microbiologist covers the smear with Kinyoun carbolfuchsin
   b. The microbiologist then decolorizes the smear with acid-alcohol; during this process most bacteria lose their red color, depending on their lipid content
   c. A brilliant green or methylene blue is applied next to stain the decolorized bacteria
3. A *flagella stain* improves the resolution of flagella by coating the specimen with a thick precipitate, or mordant, followed by staining with silver nitrate, carbolfuchsin, or methylene blue
4. A *capsule stain,* also called an India ink stain, is a negative stain for capsules
   a. To produce this stain, the microbiologist mixes India ink with the organisms
   b. If capsules are present, the India ink outlines them in brown-black
   c. A counterstain, such as methylene blue or safranin, is then applied to the bacteria; thus, the stained bacteria is surrounded by a clear area, the capsule, against a brown-black background
5. A *spore stain* identifies the presence of highly resistant **spores** or resting stages of microorganisms
   a. Spores are important to detect because they are relatively resistant to drying, sunlight, and control measures
   b. This type of stain is created by applying methylene blue, which acts as a negative stain for spores, to the smear
   c. Another technique for producing a spore stain is to apply carbolfuchsin to the spores, followed by counterstaining with nigrosin
6. A *fluorescent stain* fluoresces under ultraviolet light to enhance visualization; fluorescent dyes, such as acridine orange and auramine O, are used for positive stains of microorganisms
7. The *Wright stain* and *Giemsa stain* are used to stain blood cells and their parasites based on the pH level of the microorganism's cells
   a. These stains are a mixture of eosin (red) and methylene blue
   b. Alkaline pH areas of the cells stain blue
   c. Acid pH areas of the cells stain red
8. The *Gomori methenamine silver stain,* similar to the fluorescent stain, is used to stain yeast, fungi, and the bacteria that cause legionellosis

Study Activities

1. State three differences between prokaryotic cells and eukaryotic cells.
2. Draw a composite bacterial cell that shows the following: capsule, slime layer, cell wall, pili, and flagella.
3. Discuss the atmospheric conditions needed by aerobic, microaerophilic, anaerobic, and capnophilic organisms.
4. With the assistance of a faculty member, perform a Gram stain. Note the appearance of the bacteria on completion of the staining procedure. Would you say the bacteria are gram-negative or gram-positive? Why?

4

Growth of Microorganisms

Objectives

After studying this chapter, the reader should be able to:
- Define metabolism, respiration, gene, and enzyme.
- Describe the four phases of the population growth curve.
- Discuss anabolism and catabolism.
- Compare aerobic and anaerobic respiration.
- Explain transcription and translation of genetic information.
- Name the four cellular processes in which enzymes play a role.

I. Growth

A. General information
1. Growth is a process by which organisms increase in size or number
2. Growth in microorganisms occurs at two levels—population growth and individual growth

B. Population growth
1. Population growth refers to an increase in the numbers of an organism
   a. Population growth depends on interaction between individual organisms and their environment
   b. Population growth follows a predictable pattern, or growth curve
2. The growth curve is divided into four phases—lag phase, log phase, stationary phase, and death phase
   a. During the *lag phase,* which occurs after microorganisms are introduced into an environment, organisms do not increase in number as they adjust to their surroundings
   b. During the **log** phase, the number of organisms increases logarithmically or exponentially (for instance, 1, 2, 4, 8, 16, and so on) as they make optimum use of the resources in their environment
   c. During the *stationary phase,* the number of organisms that die equals the number of organisms being produced; thus, the population remains stable
   d. During the *death phase,* or *decline phase,* more organisms die than are produced, thereby decreasing the number of organisms; death results from a depletion of resources and a buildup of toxic waste products in the environment

C. Individual growth

1. Individual growth occurs when an organism uses energy from food to add to its cell mass
    a. Growth is recognized as an increase in size or maturity
    b. A large, mature cell can divide to produce two identical cells, each of which may carry on the cycle
2. Individual growth depends on several processes, including metabolism, production of or search for food, respiration, reproduction, and enzyme function

II. Metabolism

A. General information

1. Metabolism is the sum of the chemical processes occurring within an organism
    a. Food is produced or found and taken into the cells
    b. As food is used, energy is released
    c. The organism uses energy to convert carbon dioxide, water, nitrogen, and other simple compounds into the functioning parts of the cells
2. It is a process by which the organism produces energy for reproduction and growth
3. Metabolism comprises two opposite functions — anabolism and catabolism

B. Anabolism

1. Anabolism is a process by which simple substances, such as amino acids, are converted into more complex molecules, such as proteins
2. The anabolic process involves constructive use of the energy available through catabolism (for example, amino acids join to form peptides, and petides bond to produce the large molecules of protein)
3. Anabolism produces amino acids, proteins, lipids, and carbohydrates for use by the organism
    a. Amino acids are essential to protein production
    b. Proteins and lipids are important building blocks of structures, such as cell walls and membranes
    c. Carbohydrates are involved in the storage and release of energy

C. Catabolism

1. Catabolism is a process by which complex molecules (food), such as starch, are converted into simpler molecules, such as sugars and, finally, carbon dioxide and water
2. The breakdown of these complex molecules releases energy, which is used in anabolism
    a. For example, in starch (a carbohydrate), the breakdown releases simple sugars that are used as fuel within cells to provide energy, along with carbon dioxide and water
    b. Interestingly, the products of the catabolic reaction include components used originally in anabolism to construct the complex molecules

3. Catabolism also produces waste products that must be eliminated from the organism
   a. Acids and alcohols are produced from partial catabolism of carbohydrates
   b. Such gases as carbon dioxide and hydrogen sulfide are produced from complete catabolism of carbohydrates and sulfur compounds, respectively

III. Food

A. General information
1. All organisms must have food to ensure their growth and survival
2. Food provides energy for metabolism and carbon for building organic compounds
3. Photosynthesis and chemosynthesis are the processes by which living organisms produce food

B. Photosynthesis
1. Photosynthesis is the means used by chlorophyll-containing plants and blue-green algae to convert energy from sunlight into chemical energy
2. The light energy acts on the chlorophyll (green pigments of plant cells) to produce adenosine triphosphate (ATP)
   a. ATP is found in all cells
   b. The phosphates are held together by high-energy chemical bonds
      (1) Energy is created when the bonds are made
      (2) Energy is released when the bonds are broken
3. Photosynthesis also produces carbohydrates from the carbon dioxide and water in the chlorophyll

C. Chemosynthesis
1. Chemosynthesis is a means used by certain bacteria and algae to synthesize carbohydrates by using energy from chemical reactions rather than energy from light
2. Organisms that produce food through chemosynthesis are classified as autotrophs or heterotrophs
   a. *Autotrophs,* also called chemoautotrophs, can turn inorganic matter (matter other than plant or animal) and carbon dioxide into proteins, carbohydrates, and other cellular components; they do not need to rely on other living organisms for food (an example is the common gram-negative bacillus Pseudomonas)
   b. *Heterotrophs* cannot produce their own food and must consume organic matter produced by other organisms (examples are *saprophytes* that live on decaying organic matter and parasites that rely on a host)

IV. Respiration and related reactions

A. General information

1. Respiration and related reactions release energy by the *oxidation* of food sources
2. Respiration in microorganisms can be aerobic (using oxygen) or anaerobic (using *oxidants* other than oxygen)
3. Energy transfer occurs via electron transport and results in a series of *reduction* reactions

B. Aerobic respiration

1. Aerobic respiration converts carbohydrates to organic acids, carbon dioxide, and water
2. It provides high-energy ATP at a rate of 38 ATP molecules for each carbohydrate molecule; ATP is a critical component of many cellular reactions involving enzymes
3. The Krebs cycle is a series of reactions in which energy is generated and then stored in phosphate bonds (see *Krebs Cycle,* page 26)
   a. The breakdown of glucose (glycolysis) produces pyruvic acid
   b. Pyruvic acid reacts with Coenzyme A to produce acetyl-CoA
   c. Acetyl-CoA combines with oxaloacetic acid already in the cycle to produce a series of intermediate compounds, waste carbon dioxide, and ATP
   d. Each step is a separate enzyme reaction

C. Anaerobic reactions

1. Anaerobic metabolism converts carbohydrates to organic acids, alcohols, hydrogen, and carbon dioxide
2. It uses hydrogen, carbon, or sulfate rather than oxygen as an oxidant
3. This type of reaction provides high-energy ATP at a rate of two ATP molecules for each carbohydrate molecule
4. The Embden-Meyerhof pathway is a commonly used term to describe anaerobic metabolism (see *Embden-Meyerhof Pathway,* page 27)
   a. Glycolysis produces glyceraldehyde 3-phosphate
   b. Glyceraldehyde 3-phosphate is oxidized to intermediate compounds with ATP formation
   c. Pyruvic acid is reduced by the process of *fermentation*
   d. Products include acids, alcohols, hydrogen, and carbon dioxide

D. Energy-producing reactions in microorganisms

1. Organisms that use only aerobic respiration are called aerobes; examples include Pseudomonas
2. Organisms that use only anaerobic reactions are called anaerobes; examples include Clostridia
3. Organisms that can use both aerobic and anaerobic reactions are known as facultative anaerobes; most bacteria are facultative anaerobes

Krebs Cycle

The Krebs cycle is a series of reactions in which energy is generated and then stored in phosphate bonds.

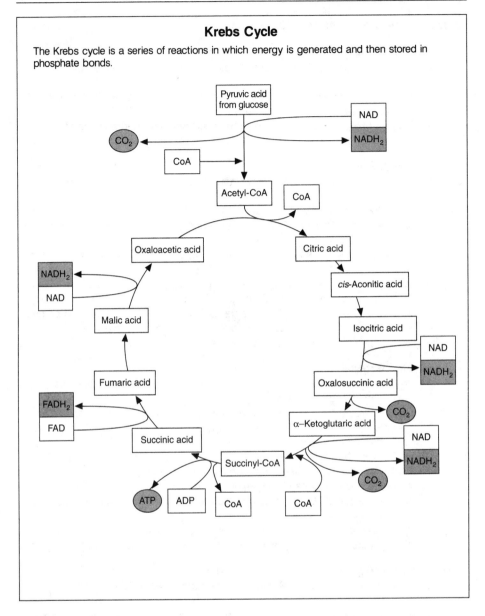

V. Reproduction

A. General information

1. Reproduction is the process of producing new individual organisms, thereby contributing to population growth
2. Through reproduction, genetic information can be transmitted asexually or sexually

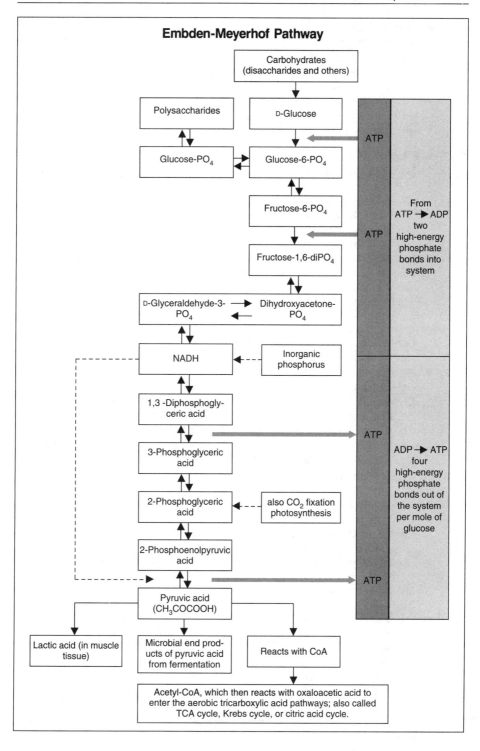

Embden-Meyerhof Pathway

Carbohydrates
(disaccharides and others)

Polysaccharides   D-Glucose

Glucose-PO$_4$ Glucose-6-PO$_4$ ATP

Fructose-6-PO$_4$

ATP

Fructose-1,6-diPO$_4$

D-Glyceraldehyde-3-PO$_4$ → Dihydroxyacetone-PO$_4$

NADH   Inorganic phosphorus

1,3 -Diphosphogly-ceric acid

3-Phosphoglyceric acid   ATP

2-Phosphoglyceric acid   also CO$_2$ fixation photosynthesis

2-Phosphoenolpyruvic acid

ATP

Pyruvic acid
(CH$_3$COCOOH)

Lactic acid (in muscle tissue)

Microbial end products of pyruvic acid from fermentation

Reacts with CoA

Acetyl-CoA, which then reacts with oxaloacetic acid to enter the aerobic tricarboxylic acid pathways; also called TCA cycle, Krebs cycle, or citric acid cycle.

From ATP → ADP two high-energy phosphate bonds into system

ADP → ATP four high-energy phosphate bonds out of the system per mole of glucose

B. Asexual reproduction
1. Asexual reproduction refers to any reproductive process that does not involve the union of gametes (reproductive cells)
2. Microorganisms such as bacteria and yeasts usually reproduce asexually by fission or budding, producing two cells with identical characteristics
   a. A mature cell (called the parent cell) produces a crosswall and divides into two equal daughter cells
   b. The daughter cells also may separate, thereby continuing the process

C. Sexual reproduction
1. Sexual reproduction in sporozoa, plants, and animals involves the union of gametes
   a. Specialized cells, called *gametocytes,* are produced
   b. Gametocytes are either male or female
   c. The union of gametocytes results in the creation of a new generation of organisms
2. Microorganisms such as bacteria and protozoa occasionally reproduce in a simple sexual process called *conjugation*
   a. During conjugation, two cells join, share genetic material, and separate
   b. The resulting cells possess the characteristics of both parent cells

D. Genetic transmission
1. Genetics is a branch of biology that is concerned with the heredity and variation of organisms
2. All hereditary traits are encoded on chromosomes and passed from one generation to the next through reproduction
3. Chromosomes are composed of deoxyribonucleic acid (DNA)
   a. DNA is composed of long chains of subunits, called nucleotides
   b. It is duplicated during reproduction by the process of replication
   c. In replication, an intact DNA chain serves as a model, or template, to which nucleotides are sequentially attached; because the original DNA serves as the template, an exact copy is produced to pass on during reproduction
4. *Genes,* the biologic units of heredity, are located at a particular position on each chromosome
   a. For example, in humans, the gene that controls the composition of white blood cell membranes is located on chromosome number 6
   b. Genes carry the code for specific protein products, such as enzymes and structural components
5. The genetic code is decoded, or read, during transcription

E. Transcription and translation
1. *Transcription* is an enzymatic process during which the genetic information contained in the gene's DNA is used to assemble a ribonucleic acid (RNA) molecule
   a. This copy is called messenger RNA (mRNA)
   b. The mRNA attaches to a ribosome, which is the site of protein synthesis

2. *Translation* is a process during which the genetic information contained in the mRNA dictates a specific sequence of amino acids; this sequence results in a protein molecule
3. Transcription and translation are controlled by inducers and suppressors, which are themselves under gene control
   a. Inducers turn on the processes
   b. Suppressors turn off the processes

F. Alteration in genetic transmission
1. The transmission of genetic information can be affected by mutation, transformation, transduction, or recombination
2. *Mutations* are random changes in the genetic code resulting from an error in replication, breakage or partial deletion of the chromosome, or insertion of a DNA fragment at a new point in the genes
   a. Mutations can occur spontaneously or can be produced in a laboratory with chemicals, ultraviolet light, or radiation
   b. They are passed on during reproduction and effect permanent and heritable changes in the genes
   c. The new characteristics they introduce into the genes may account for the diversity of life
3. *Transformation* is the exchange of a DNA fragment between cells
   a. For example, one microorganism may acquire a characteristic from another organism
   b. Resistance to antibiotics might be passed from one bacterium to another
4. *Transduction* is the exchange of DNA between cells by a virus, which infects one cell and then another, carrying the DNA to the second cell; effects are similar to those of transformation
5. *Recombination* is the linking of genes in a new order
   a. Microbiologists can create new cellular products by taking advantage of the genes' ability to recombine
   b. For instance, a gene from the hepatitis virus can be recombined in the yeast Saccharomyces to produce an effective vaccine against hepatitis
   c. Gene-recombinant technology is a rapidly developing field in microbiology and medicine

VI. Enzyme Function

A. General information
1. *Enzymes* are protein catalysts that speed up chemical reactions without being destroyed or altered during the process
2. Enzymes turn a specific **substrate** (substance acted on by an enzyme) into a product (substance that results from enzyme action)
3. To affect a chemical reaction, enzymes may require the help of cofactors called coenzymes
4. Enzymes are involved in the cellular processes of growth, metabolism, respiration, and reproduction

B. Naming enzymes

1. Enzymes are named with the suffix *-ase* for the reaction that they catalyze
2. The name is determined by the substrate acted on and by the result obtained
   a. Deaminase acts on an amino compound to produce ammonia
   b. Decarboxylase accelerates decarboxylation (removal of carbon dioxide) from amino acids
   c. Dehydrogenase accelerates the removal of hydrogen from metabolites (substances essential to metabolism)
   d. Catalase acts on hydrogen peroxide to produce free oxygen
   e. Hydrolase accelerates the breakdown of a large molecule into two or more smaller molecules with the addition of a water molecule
   f. Kinase converts a proenzyme (an inactive precursor of an enzyme) to an active enzyme
   g. Lipase acts on fat (lipid) molecules, splitting them into glycerol and fatty acids
   h. Oxidase catalyzes oxidation

C. Coenzymes

1. Coenzymes, which participate in reactions with enzymes, are usually composed of phosphorus or vitamins
2. Enzymes and coenzymes must unite in order to function
3. Coenzymes store or release energy and accept the oxidation products of the chemical reactions
4. Common coenzymes are ATP, nicotinamide-adenine dinucleotide (NAD), and flavin-adenine dinucleotide (FAD)

D. Classifying enzymes

1. The enzymes found in microorganisms are classified according to when and where they function
2. *Constitutive* enzymes are always present as an integral part of a cell; dismutase, for example, allows aerobes to grow in the presence of oxygen
3. *Adaptive* enzymes, also called induced enzymes, are produced only in response to the existence of a specific substrate; cytochrome oxidase, for example, is produced only in the presence of iron
4. *Exoenzymes* are produced in a cell but function outside the cell
   a. Many of a pathogen's toxic effects are caused by exoenzymes
   b. Examples include hemolysins, which lyse red blood cells, and coagulases, which clot protein

---

Study Activities

1. Describe the ways in which individual growth depends on metabolism, respiration, and reproduction.
2. Research the role played by photosynthesis and chemosynthesis in food production. Then lead a class discussion on your findings.
3. Distinguish between autotrophs and heterotrophs.
4. List the functions of five enzymes and five coenzymes.

5

Identifying Microorganisms

Objectives

After studying this chapter, the reader should be able to:
• Define nomenclature and taxonomy.
• Identify the four characteristics used to assign a genus and species name.
• Describe at least eight characteristics used to identify a microorganism.

I. Nomenclature

A. General information

1. The process of assigning names to organisms, or nomenclature, follows a formalized set of rules
    a. The first step is identifying and agreeing on the organism's characteristics; two U.S. organizations that carry out this task are the American Society for Microbiology and the American Type Culture Committee
    b. Based on its identifying characteristics, the organism is named
    c. Properly named organisms are grouped into categories based on their similarities
2. *Taxonomy* is the systematic classification of organisms based on their categories, or taxa; this classification can be used to compare similar or disparate organisms (see *Taxonomy of Humans and Escherichia coli,* page 32)
    a. A *species,* the most specific category, comprises all organisms with characteristics similar enough to be considered of one and the same type
    b. A *genus* comprises all related species
    c. A *tribe* comprises all related genera
    d. A *family* comprises all related tribes
    e. An *order* comprises all related families
    f. A *class* comprises all related orders
    g. A *phylum* comprises all related classes
    h. A *kingdom,* the broadest category, comprises all related phyla

B. Binomial nomenclature

1. *Binomial* (two-name) *nomenclature* is the use of genus and species names to identify organisms
2. Historians credit Carolus Linnaeus, an 18th-century Swedish naturalist, with its development

Taxonomy of Humans and *Escherichia coli*

The chart below can be used to compare the taxa for such disparate organisms as humans and the gram-negative bacterium *E. coli*. Note that the taxonomy progresses from more general and inclusive terms to more specific and exclusive ones.

TAXONOMY	HUMANS	*ESCHERICHIA COLI*
Kingdom	Animalia	Protista (Monera)
Phylum	Chordata	Schizophyta (Protophyta)
Class	Mammalia	Schizomycetes
Order	Primates	Eubacteriales
Family	Hominidae	Enterobacteriaceae
Tribe	None	Escherichieae
Genus	Homo	Escherichia
Species	sapiens	coli

    a. Before Linnaeus's work, organisms were identified by polynomial (many-name) nomenclature, which entailed using a series of names to describe a single organism

    b. The innovation of binomial nomenclature greatly simplified the naming of organisms

II. Genus and Species Names

A. General information

    1. Genus and species names are latinized terms that describe specific information about an organism

    2. An organism is assigned a name based on its identifying characteristics, including historical information, physical features, culture requirements, and diseases produced

    3. Certain conventions are followed when assigning genus and species names

        a. The genus name is always given first

        b. The species name follows the genus name

        c. New names are accepted internationally only after other scientists can comment on and verify the classification

B. Naming an organism

    1. An organism may be assigned a name based on historical information, such as the name of the organism's discoverer, its geographic location when it was discovered, or the site that it typically inhabits

        a. The bacterium *Escherichia coli*, for instance, derives its genus name from microbiologist T. Escherich

        b. Its species name is derived from the large intestine, or colon, which the bacterium usually inhabits

2. Physical features, such as size, shape, color, or odor, may be used to develop a genus and species name
   a. *Staphylococcus aureus* is named for its organization into grapelike clusters of gram-positive cocci and its golden yellow color
   b. *Streptococcus malodoratus* is named for its organization in chains of cocci and for the foul odor it produces
3. An organism's culture requirements—such as the nutrients, atmosphere, temperature, and metabolic products needed to grow it—may be the basis for its name
   a. *Haemophilus aegyptius* is named for its requirement for blood (heme) and its discovery in Egypt
   b. *Nitrasomonas* is named for its nitrogen requirement
4. An organism can be named for the disease that it causes
   a. *Propionibacterium acnes* is so named because it produces propionic acid and causes acne
   b. A name can persist long after the organism's association with a particular disease has been disproven; for example, *Listeria monocytogenes* is no longer considered a cause of infectious mononucleosis, yet its name remains unchanged

III. Procedures Used to Identify Microorganisms

A. General information
1. To ensure proper identification, scientists must thoroughly evaluate a microorganism
2. This evaluation includes examining the organism under a microscope to identify its characteristics and determine its reaction to stains; growing the microorganism in pure culture to assess its growth characteristics; and assessing other characteristics, such as color and odor (see *Identifying Streptococcus pyogenes,* page 34)

B. Microscopic examination
1. Microscopic examination allows the microbiologist to evaluate several of the organism's identifying characteristics
   a. Morphology—bacilli, cocci, or spirochetes—is apparent
   b. Organization—single, paired, clustered, or chained—can be determined
   c. Motility—nonmotile or motile by means of one or more flagella, pseudopods, or *cilia*—can be described
   d. Specific features—a capsule or flagella—can be seen
2. After an organism has been stained, microscopic examination can reveal the stain reaction (for example, gram-positive, gram-negative, acid-fast positive, or acid-fast negative)

C. Culturing
1. Growing an organism in culture can help determine atmospheric requirements—aerobic, anaerobic, facultative, microaerophilic, or capnophilic
2. Culturing can reveal an organism's temperature requirements; that is, whether it is cryophilic, mesophilic, or thermophilic

---

Identifying *Streptococcus pyogenes*

The following chart lists the steps that a microbiologist would take to identify *S. pyogenes*, the cause of a common infection.

STEPS	TYPICAL RESULTS
Determine the source	Infected throat
Culture	On blood agar and PEA agar, organism is aerobic and facultative with mature colonies in 18 to 24 hours at 37° C
Stain for morphology	Gram-positive cocci in chains
Note characteristics	Beta-hemolytic on blood agar Highly susceptible to low doses of the antibiotic bacitracin
Detect antigens	Antigens that react with antibodies to group A *Streptococcus*
Make identification	Group A *Streptococcus pyogenes*

---

3. The procedure also can help microbiologists assess speed and type of reproduction
   a. Some microorganisms are fully developed in 18 to 24 hours, whereas others require 4 to 8 weeks to reach maturity
   b. A microorganism can reproduce through fission, budding, conjugation, or sexual reproduction

D. Other identifying characteristics
1. An organism's color or appearance—white, gray, yellow, green, blue-green, brown, red, orange, purple, black, or fluorescent—is used to identify it
2. Organisms may possess a distinct odor—foul, sweet, fruity, or butterlike
3. *Hemolysis*—the degree to which an organism destroys red blood cells—is another identifying characteristic; hemolysis is classified as complete (beta), partial (alpha), or none (gamma)
4. Organisms can be classified according to acid production; for example, certain organisms produce acetic acid, butyric acid, lactic acid, or propionic acid from carbohydrates and alcohols
5. The type of gas produced by a microorganism—for example, ammonia, carbon dioxide, or hydrogen sulfide—is an identifying characteristic
6. An organism's enzyme production and whether it is adaptive or constitutive for metabolism is another characteristic evaluated by the microbiologist
7. How an organism utilizes specific nutrients—such as citrate, arginine, lysine, ornithine, and phenylalanine—is important to its classification
8. A microbiologist evaluates whether an organism is susceptible or resistant to drying, antiseptics, sterilization, or *antimicrobial* agents
9. Evaluating an organism's relationship with other organisms—commensal, mutual, parasitic, or pathogenic—is part of identifying its characteristics

10. A microbiologist attempts to detect specific antigens and compares the structural components of an organism's cells through antigen-antibody reactions
    a. A specific *antigen* is unique to a particular organism
    b. The antigen will react with its specific antibody
    c. An antibody of a known type can be mixed with an unknown antigen source
       (1) If a reaction occurs, the source must contain the appropriate antigen (positive identification)
       (2) No reaction means that the antigen is lacking (no identification)
11. *DNA probes* are used to compare similarities in the chromosomes of different microorganisms

E. Numerical identification and computerization
   1. After the microbiologist identifies an organism's characteristics, selected characteristics are assigned numbers
   2. The sum of the weighted characteristics is compared with sums from known organisms
   3. This comparison is used to identify the microorganism
      a. Ideally, only one organism will show a pattern of characteristics that yield a particular number
      b. This number becomes a shorthand notation for one set of characteristics
      c. For example, if *Escherichia coli* yields a particular number, then any other organism that generates the same number will be presumptively identified as *E. coli*
   4. The numerical identification techniques can be computerized for rapid processing
      a. Values for the characteristics are determined by the microbiologist and entered into a computer or read directly by a computer equipped with an optical scanner
      b. The computer processes the values, matching them with one identification or (if no perfect matches result) giving several possible identifications with probabilities for each and additional results that will clarify the uncertainty

Study Activities

1. Consult a textbook for a key to the identification of microorganisms. Then describe the steps one must take and the specific identifying characteristics one must consider to arrive at a conclusion.
2. Research the contributions of Carolus Linnaeus to biology.
3. Determine the characteristics of and attempt to identify a microorganism selected by your instructor. Confirm your identification with the instructor.
4. Repeat the previous activity, only this time using a rapid means of identification provided by your instructor.

6

Bacteria

Objectives

After studying this chapter, the reader should be able to:
- List six ways of classifying microorganisms.
- Describe the primary characteristics of gram-positive and gram-negative cocci.
- Distinguish among gram-positive, gram-negative, and gram-variable bacilli.
- Describe the primary characteristics of spirochetes.
- Name at least one member of each of the six groups of bacteria listed above.

I. Bacteriology

A. General information
1. Bacteriology is the study of bacteria, a diverse group of microorganisms
2. Bacteria are widely dispersed throughout nature; they can be found beneath the ice of Antarctica as well as in hot springs and oceanic thermal vents
3. These microorganisms range from 1 to 8 micrometers in length and less than 1 micrometer in diameter
4. Bacteria are prokaryotic; that is, their cells do not contain a true nucleus
5. Bacteria play an important role in interactions with the environment and higher life forms
   a. Bacteria are at the bottom of food chains in many biological systems
   b. Bacteria live in, on, and around other microorganisms, plants, and animals, as commensals, mutualists, and parasites
   c. Humans use bacteria in diverse ways, from the making of paper and medicines to the clean-up of oil spills
6. Although commonly part of the normal *flora* of plants, animals, and humans, bacteria can be pathogenic
7. Bacteria are classified according to their morphology, stain reactions, atmospheric requirements, organization, physical and biochemical composition, and antigenic structure

B. Classifications
1. Bacteria can be classified by their morphology, or shape
   a. Some bacteria are shaped as spheres (cocci)
   b. Some bacteria are shaped as rods (bacilli)
   c. Other bacteria are shaped as spirals (spirochetes)

2. Gram-stain reactions can be used to distinguish common bacteria
   a. Bacteria are classified as gram-positive, gram-negative, or gram-variable, depending on the age of the cells
   b. Bacteria that do not stain readily by Gram stain are detected by other techniques, such as darkfield microscopy or the use of special stains
3. Atmospheric requirements—aerobic, microaerophilic, anaerobic, facultative, or capnophilic—are another means of classifying bacteria
4. Bacteria also can be identified by their organization (for example, single, pairs, or clusters), color, and the type of medium on which they grow
5. The primary groups of bacteria are the gram-positive cocci, gram-negative cocci, gram-positive bacilli, gram-negative bacilli, gram-variable bacilli, and spirochetes

II. Gram-Positive Cocci

A. General information
1. Gram-positive cocci comprise aerobes, anaerobes, and facultative organisms
2. They are organized in singles, pairs, and clusters, as well as short and long chains

B. Means of identification
1. Gram-positive cocci can be identified through biochemical tests, hemolysis assessment, and serologic methods (see *Identifying Medically Important Gram-Positive Cocci,* page 38)
2. Biochemical tests for gram-positive cocci include catalase, deoxyribonuclease, coagulase, salt tolerance, and resistance to bile salts and antimicrobial agents
   a. In the catalase test, hydrogen peroxide is broken down into water and oxygen
   b. In the deoxyribonuclease test, DNA is degraded
   c. In the coagulase test, human or rabbit plasma clots
   d. In the salt tolerance test, bacterial growth occurs in concentrations of sodium chloride up to 6.5%
   e. Resistance to bile salts and antimicrobial agents is confirmed when bacteria grow in media containing bile salts, bacitracin, or optochin
3. Hemolysis (destruction of red blood cells) on blood agar is classified as complete (beta), partial (alpha), or none (gamma)
   a. Typically, the more complete the hemolysis, the more pathogenic the cocci
   b. *Staphylococcus aureus* and *Streptococcus pyogenes,* which are beta-hemolytic, are highly pathogenic
4. Serologic methods are used to determine specific antigens within the microorganism's cells
   a. Streptococci are grouped into 15 serotypes based on the serologic findings
   b. Group A streptococci are the most pathogenic serotype

C. Location
1. Gram-positive cocci are widespread throughout the environment

Identifying Medically Important Gram-Positive Cocci

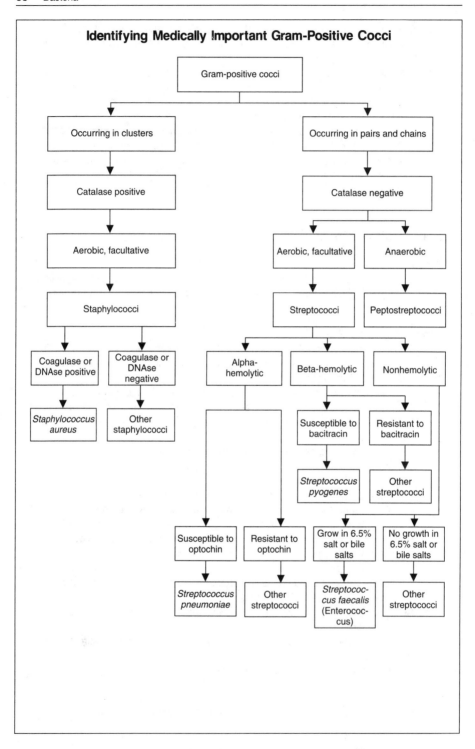

    2. They are found on human skin and hair and as normal flora in the human upper respiratory tract, gastrointestinal tract, and mucous membranes

    3. These microorganisms can be isolated from wound infections and infections of the respiratory tract, urinary tract, skin, and blood

D. Importance

    1. Gram-positive cocci include human pathogens and opportunists, such as *Staphylococcus aureus, Staphylococcus hominis, Streptococcus pyogenes, Streptococcus viridans, Streptococcus pneumoniae,* and group D streptococci (enterococci)

       a. *Staphylococcus aureus* is a cause of food poisoning, skin infections, and **sepsis**

       b. *Streptococcus pyogenes* causes strep throat and scarlet fever

       c. *Streptococcus pneumoniae* is one of the primary causes of bacterial meningitis

    2. These organisms are often highly resistant to treatment with common antimicrobial agents

       a. Penicillinase inactivates penicillin and cephalosporins

       b. Penicillinase-resistant antibiotics or vancomycin must be used instead

E. Taxonomic classification: Family Micrococcaceae

    1. Gram-positive cocci in the family Micrococcaceae demonstrate aerobic and facultative respiration

    2. These microorganisms, which can occur in singles, pairs, or clusters, are white to golden-yellow

    3. Members of this group grow on nutrient, enriched, and gram-positive selective media

    4. Examples include the human pathogen *Staphylococcus aureus* and normal flora of the skin

F. Taxonomic classification: Family Peptococcaceae

    1. The microorganisms in this family demonstrate anaerobic respiration

    2. They occur in short to long chains and have a gray-white color

    3. Members of this family grow on enriched media in anaerobic conditions

    4. One example is the normal human flora *Peptostreptococcus,* an opportunist

G. Taxonomic classification: Family Streptococcaceae

    1. These microorganisms exhibit aerobic, microaerophilic, and facultative growth, actively fermenting in all conditions

    2. They occur in short to long chains and can be colorless or have a gray-white pigmentation

    3. They grow on nutrient, enriched, and gram-positive selective media

    4. A member of this family is the human pathogen *Streptococcus pyogenes*

III. Gram-Negative Cocci

A. General information

    1. Gram-negative cocci comprise aerobes, anaerobes, and capnophiles

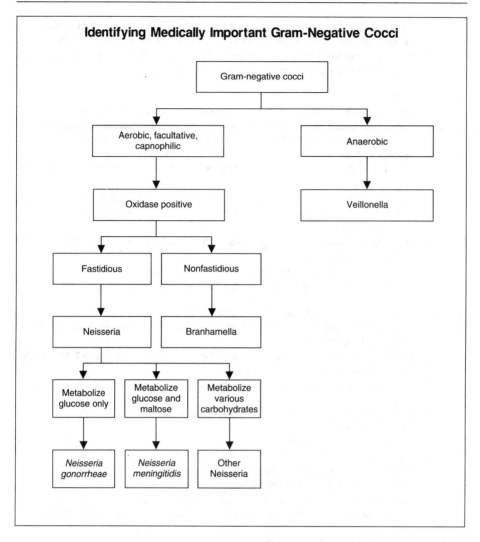

Identifying Medically Important Gram-Negative Cocci

2. They can be found as singles, pairs, or short chains

B. Means of identification
1. Gram-negative cocci can be identified through biochemical and serologic tests
   (see *Identifying Medically Important Gram-Negative Cocci*)
2. Biochemical tests for gram-negative organisms include oxidase and
   carbohydrate utilization
   a. In the presence of oxidase reagent, bacterial colonies producing the
      enzyme turn purple to black
   b. Yellow indicates acid production from carbohydrates, such as glucose,
      sucrose, and maltose
3. Serologic tests are conducted for specific antigens to identify these organisms

C. Location

1. Gram-negative cocci are part of the normal human flora
2. They can be isolated from infections of the genital tract, blood, cerebrospinal fluid, and joints

D. Importance

1. The gram-negative cocci *Neisseria gonorrhoeae* is the cause of an increasing number of cases of gonorrhea
2. *Neisseria meningitidis* type B causes approximately 1,500 cases of meningitis in the United States each year
3. Gram-negative cocci are increasingly resistant to treatment with common antimicrobial agents

E. Taxonomic classification: Family Neisseriaceae

1. The microorganisms in this family are obligate aerobes with an affinity for carbon dioxide (that is, they are capnophilic)
2. They occur in singles and pairs and are either colorless or have gray-to-buff pigmentation
3. They can be grown on enriched and selective media, such as chocolate agar or Thayer-Martin agar
4. Members of this group include the intracellular parasite *Neisseria gonorrhoeae*, other *Neisseria* species, *Moraxella* (formerly *Branhamella*) species, and the normal flora of the mouth and upper respiratory tract

F. Taxonomic classification: Family Veillonellaceae

1. These microorganisms demonstrate obligate anaerobic respiration
2. They occur in pairs and short chains and have a gray-white color
3. They grow on enriched media under anaerobic conditions
4. A member of this family is the normal human flora *Veillonella*

IV. Gram-Positive Bacilli

A. General information

1. Gram-positive bacilli comprise aerobes and anaerobes, as well as facultative and capnophilic organisms
2. They are organized singly, in pairs, or in short to long chains

B. Means of identification

1. Gram-positive bacilli can be identified by their rate of growth, pigmentation, and motility (see *Identifying Medically Important Gram-Positive Bacilli,* page 42)
2. Biochemical tests used to identify these microorganisms include catalase, carbohydrate utilization, niacin production, and tellurite reduction
   a. In the niacin test, a yellow color change in the niacin reaction confirms production of this vitamin
   b. In the tellurite reduction test, media that contain tellurite turn gray to black if bacteria are capable of reducing this chemical

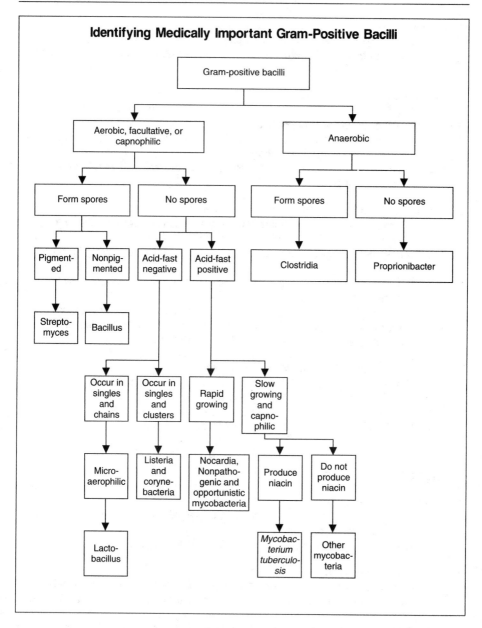

Identifying Medically Important Gram-Positive Bacilli

3. Gram-positive bacilli also can be identified by their products of fermentation, such as acids and gas from carbohydrates

C. Location
1. Gram-positive bacilli are widely distributed in soil and water
2. They also can be found in, on, or around plants and animals, including humans

3. These organisms can be isolated from infections of the skin, blood, lungs, bones, and urinary tract, as well as from wounds and bites

D. Importance
1. Gram-positive bacilli cause various infections, including tetanus, gas gangrene, and the nearly 25,000 cases of tuberculosis diagnosed each year in the United States
2. The *Bacillus, Clostridium,* and *Listeria* species also cause food poisoning
3. As normal human flora, gram-positive bacilli act to limit or prevent the overgrowth of pathogens

E. Taxonomic classification: Family Corynebacteriaceae
1. The organisms in this family exhibit aerobic respiration
2. They occur as singles, pairs, and irregular clusters (Chinese letter shapes)
3. They can be identified by their gray-white color, and they grow on nutrient and enriched media
4. *Corynebacterium diphtheriae* (the cause of diphtheria) belongs to this family

F. Taxonomic classification: Family Mycobacteriaceae
1. These gram-positive bacilli demonstrate aerobic and capnophilic respiration
2. They occur as singles and multiple chains (cords)
3. An identifying characteristic is their buff, yellow, or orange-red color
   a. Pigmentation occasionally depends on light
   b. For example, *Mycobacterium kansasii* is buff colored in the dark but turns bright yellow when exposed to light
4. Mycobacteria can be grown on enriched media (egg based or synthetically defined)
   a. It can take as long as 6 weeks to produce colonies
   b. When grown on culture, *Mycobacterium tuberculosis* produces niacin, an important aid in identification
5. Members of this family are acid-fast positive when stained
6. *Mycobacterium tuberculosis* (the cause of tuberculosis), *M. leprae* (the cause of leprosy), and *M. bovis* (the cause of bovine tuberculosis) belong to this family

G. Taxonomic classification: Family Nocardiaceae
1. These organisms demonstrate aerobic respiration
2. They can be seen as singles, chains, or branching forms
3. Buff to yellow in color, these organisms grow on nutrient and enriched media, including media selective for mycobacteria and fungi
4. When stained, members of this family are partially acid-fast positive
5. *Nocardia asteroides* (the cause of the respiratory tract infection nocardiosis) is included in the family Nocardiaceae

H. Taxomonic classification: Family Propionibactereaceae
1. Anaerobic respiration is demonstrated by these gram-positive bacilli
2. Occurring as singles, these organisms have a gray-white color
3. They can be grown on enriched media under anaerobic conditions
4. *Propionibacterium acnes* (a cause of acne) is a member of this family

I. **Taxonomic classification: Family Streptomycetaceae**
   1. Aerobic respiration is a characteristic of the members of this family
   2. These branching-chain organisms contain spores
      a. Spores, the resting stage of a bacterial cell, are resistant to environmental changes
      b. They are produced as a means of withstanding adverse conditions, such as drying or lack of nutrients
      c. When conditions are favorable, spores germinate and start growing again
   3. These organisms can be buff to yellow, orange, red, gray, or black
   4. They grow on nutrient and enriched media and give off an earthy odor
   5. This family includes *Streptomyces* species, the source of the antibiotic streptomycin

J. **Taxonomic classification: Genus *Bacillus***
   1. Members of the genus *Bacillus* demonstrate aerobic respiration
   2. They occur as short to long chains of spore-producing bacilli
   3. *Bacillus* species are white to gray in color and grow on nutrient and enriched media
   4. Members of this genus include the plant pathogen *Bacillus cereus* and the animal and human pathogen *Bacillus anthracis* (the cause of anthrax)

K. **Taxonomic classification: Genus *Clostridium***
   1. These microorganisms demonstrate anaerobic respiration
   2. They are found as single spore-producing bacilli
   3. Gray-white in color, they are markedly hemolytic on blood agar
   4. Clostridia grow on enriched media in anaerobic conditions
   5. Members of this genus include *Clostridium tetani* (the cause of tetanus), *C. perfringens* (the cause of gas gangrene), and *C. botulinum* (the cause of botulism)

L. **Taxonomic classification: Genus *Lactobacillus***
   1. Active fermentation is a characteristic of the members of this genus
   2. Occurring as singles and chains, lactobacilli are pigmented gray-white
   3. They grow on nutrient and enriched media at a pH of 6
   4. This genus includes *Lactobacillus acidophilus,* part of the normal human flora of the gastrointestinal and genital tracts

M. **Taxonomic classification: Genus *Listeria***
   1. Members of this genus demonstrate aerobic respiration
   2. They are organized as singles and have gray to black pigmentation
   3. They grow on enriched and selective media, even at temperatures of 2° to 15° C
   4. A member of this genus is *Listeria monocytogenes,* which is found in many animals and as an opportunistic human pathogen; it is also a cause of food poisoning

V. Gram-Negative Bacilli

A. General information
1. Gram-negative bacilli comprise aerobes, anaerobes, and facultative organisms
2. They are organized as singles, pairs, sausage-like masses, short to long chains, and *filamentous* forms

B. Means of identification
1. Several means exist for identifying gram-negative bacilli (see *Identifying Medically Important Gram-Negative Bacilli,* page 46)
2. Pigmentation and motility are two characteristics useful in identification
3. Biochemical tests include carbohydrate oxidation and fermentation, amino acid metabolism, oxidase, and other specific enzyme analyses
   a. In the amino acid metabolism test, decarboxylation and deamination of amino acids cause color changes in the media
   b. Other specific enzyme analyses include urease and gelatinase
4. Serologic determination of specific antigens can be used to determine the serotypes of *Salmonella* species, *Shigella* species, and *Haemophilus* species
5. Rapid automated methods and computerization also are being used to identify gram-negative bacilli
   a. Many advances in speed of identification and miniaturization of culture containers and biochemical tests result directly from space program technology
   b. Computerization provides access to data from thousands of tests on known microorganisms, thereby facilitating the identification of many more species

C. Location
1. Widespread throughout nature, gram-negative bacilli include human normal flora, pathogens, and opportunists, as well as pathogens of plants and animals
2. They can be isolated from infections of the urinary tract, gastrointestinal tract, respiratory tract, and blood, as well as from infected wounds

D. Importance
1. Gram-negative bacilli play a role in the decay of organic matter and *nitrogen fixation* in soil; nitrogen and nitrogen-containing waste products are chemically altered (fixed) into nutrients needed by plants
2. The gram-negative bacilli *Salmonella, Shigella,* and *Campylobacter* species are common causes of food poisoning

E. Taxonomic classification: Family Bacteroidaceae
1. Anaerobic respiration is a characteristic of this family
2. Other identifying characteristics are single organization and gray-white color
3. These organisms can be grown on enriched media under anaerobic conditions
4. *Bacteroides fragilis,* which is both normal flora in the intestinal tract and an opportunistic pathogen, is a member of this family

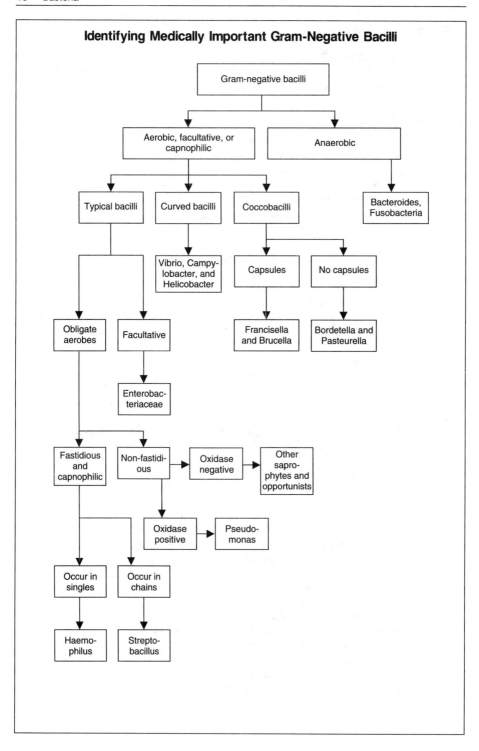

Identifying Medically Important Gram-Negative Bacilli

F. Taxonomic classification: Family Enterobacteriaceae
1. Members of this family demonstrate aerobic and facultative growth
2. They can be identified as singles and occasionally sausage-like masses
3. Gray, yellow, and red in appearance, these microorganisms grow on nutrient, enriched, and gram-negative selective media
4. Members of this family include *Escherichia coli, Salmonella* (the cause of typhoid fever), *Shigella* (the cause of dysentery), and *Yersinia pestis* (the cause of plague)

G. Taxonomic classification: Family Nitrobacteraceae
1. Characteristics of this family are aerobic respiration and organization as singles
2. These organisms are not commonly cultured because they are of interest primarily to agricultural microbiologists
3. Included in this group are important nitrogen-fixing and metabolizing genera, such as *Nitrobacter* and *Nitrosomonas*

H. Taxonomic classification: Family Pseudomonadaceae
1. Aerobic respiration characterizes this family
2. These organisms, metallic gray to green and brown, appear as singles
3. They grow on nutrient, enriched, and gram-negative selective media and produce a fruity odor
4. The animal and human pathogens *Pseudomonas aeruginosa* and *P. mallei* belong to this family; they tend to be highly resistant to many antimicrobials

I. Taxonomic classification: Family Vibrionaceae
1. These microorganisms demonstrate aerobic and facultative respiration
2. They can appear as singles or pairs of curved bacilli
3. Gray-white to purple in color, they can be grown on enriched and gram-negative selective media
4. This family includes *Vibrio cholerae* (the cause of cholera), *Campylobacter* species (the cause of acute gastrointestinal disease), and *Helicobacter* species (associated with peptic ulcers)

J. Taxonomic classification: Genus *Bordetella*
1. *Bordetella* demonstrate aerobic respiration
2. They occur as single coccobacilli and are pearl-like to metallic in color
3. Enriched medium containing fresh blood is the appropriate culture medium
4. This genus includes *Bordetella pertussis* (the cause of whooping cough)

K. Taxonomic classification: Genus *Brucella*
1. These organisms demonstrate aerobic and sometimes capnophilic respiration
2. They appear as single, short bacilli (coccobacilli)
3. These gray-white organisms grow slowly on enriched media
4. The animal and human pathogen *Brucella abortus* (the cause of brucellosis) belongs to this genus

L. Taxonomic classification: Genus *Francisella*
1. These organisms exhibit aerobic respiration

2. They occur as single, encapsulated coccobacilli and are not usually cultured because they are **fastidious** in their nutritional requirements and highly infectious
3. The animal and human pathogen *Francisella tularensis* (the cause of tularemia) belongs to this genus

M. **Taxonomic classification: Genus *Haemophilus***
   1. Aerobic and capnophilic respiration are demonstrated by these organisms
   2. They are organized as singles that may be coccobacillary to filamentous in shape
   3. They can be identified by their glistening gray color
   4. *Haemophilus* species grow on enriched media containing nicotinamide adenine dinucleotide (NAD) or **hemin**
   5. *Haemophilus influenzae* type B (a cause of meningitis) is included in this genus

N. **Taxonomic classification: Genus *Pasteurella***
   1. These single coccobacilli exhibit aerobic respiration
   2. Members of this genus are not commonly cultured because they are highly infectious
   3. An example of this genus is the animal and human pathogen *Pasteurella multocida*, which is associated with a bite or scratch from a cat or dog

O. **Taxonomic classification: Genus *Streptobacillus***
   1. Streptobacilli demonstrate aerobic respiration
   2. They occur as short to long chains and are pigmented gray-white
   3. They are grown on media enriched with serum or other body fluids
   4. An example of this genus is *Streptobacillus moniliformis* (a cause of rat bite fever)

VI. Gram-Variable Bacilli

A. **General information**
   1. Gram-variable bacilli are aerobic organisms that may stain both gram-positive and gram-negative in one smear
   2. They are organized as single coccobacilli or as filamentous forms

B. **Means of identification**
   1. Biochemical tests used to identify gram-variable bacilli include carbohydrate utilization, catalase, galactosidase activity, and growth on starch
      a. In the galactosidase test, a color change in the medium indicates presence of the enzyme
      b. In the starch test, a yellow color in the medium indicates starch metabolism
   2. Serologic determination of specific antigens can be used to differentiate gram-variable bacilli from gram-negative bacilli of similar morphology

C. Location

1. Gram-variable bacilli are opportunistic organisms found in the genitourinary tract of humans
2. These bacteria also have been found in the blood of newborns and their mothers

D. Importance

1. The gram-variable organism *Gardnerella vaginalis* is associated with bacterial vaginosis, a condition that afflicts up to 50% of females in the United States
2. Large numbers of *Gardnerella* and other bacteria overgrow the normal anaerobic flora of the vagina

E. Taxonomic classification: Genus *Gardnerella*

1. These gray-white microorganisms exhibit aerobic respiration
2. They are grown on enriched media
3. This genus consists of a single species, *Gardnerella vaginalis*

VII. Spirochetes

A. General information

1. Spirochetes are aerobic organisms
2. They occur as single, spiral forms

B. Means of identification

1. Spirochetes can be identified by their corkscrew motility and distinctive spiral morphology
2. Serologic assays can be used to determine nonspecific and specific antigens
   a. Screening tests for syphilis, for example, are nonspecific
   b. Confirmatory tests for syphilis, however, are specific for the causative agent *Treponema pallidum*

C. Location

1. Spirochetes are widespread
2. They can be isolated from infections of the genitourinary tract, skin, and blood

D. Importance

1. Spirochetes are part of the normal flora of the intestinal tracts of animals and humans
2. This group also includes pathogenic organisms that cause infections, such as syphilis and Lyme disease

E. Taxonomic classification: Family Spirillaceae

1. Members of this family are organized as singles, although they frequently form chains of spirally twisted cells
2. These organisms are grown for study by inoculation into laboratory animals
3. *Spirillum minus* (a cause of rat bite fever) belongs to this family

F. Taxonomic classification: Family Spirochaetaceae

1. Occurring as singles, these organisms are not commonly cultured because of their fastidious nature and the risk of infection
2. Examples of this family are *Treponema pallidum* (the cause of syphilis) and *Borrelia burgdorferi* (the cause of Lyme disease)

G. Taxonomic classification: Genus *Bdellovibrio*

1. These organisms are parasites of other bacteria, such as *Pseudomonas* and *Salmonella*
   a. They reproduce between the cell wall and plasma membrane of the host bacterium
   b. Like viruses, the *Bdellovibrios* leave the dead host cells to infect new bacteria
2. Bdellovibrio occur as singles and segmented forms
3. These organisms are cultured on bacterial cells growing in nutrient and enriched media

---

Study Activities

1. With your instructor, look at Gram stains of positive and negative cocci and bacilli. Describe your findings.
2. Perform a throat culture, a hand culture, a culture of air from the classroom, or a culture of water from the water fountain. How many different types of bacteria did you find?
3. Interview a health professional on the increase in sexually transmitted diseases.
4. With your instructor, perform selected tests on bacterial cultures as a means of identification.
5. Research and report on the microbiology of sewage disposal.
6. Research and report on the microbiology of soil.

7

Intermediate Organisms

Objectives

After studying this chapter, the reader should be able to:
- Define intermediate organism, L-form, and energy-dependent organism.
- List at least two identifying characteristics of forms lacking cell walls and obligate intracellular parasites.
- Discuss the importance of members of the family Rickettsiaceae and those of the family Chlamydiaceae.

I. Overview

A. General information
1. Intermediate organisms are larger and more complex than viruses but smaller and simpler than bacteria
2. They differ from bacteria and viruses in some important aspects
    a. The *L-form,* also called the L-phase variant, is a name derived from the Lister Institute in London, England, where these organisms were first studied
    b. An L-form is an intermediate organism that lacks a cell wall but can multiply in a hypertonic medium or in a host
    c. Obligate intracellular parasites reproduce only inside host cells
3. Intermediate organisms include a variety of animal and human parasites
    a. They are less than 2 micrometers in size
    b. Most of these organisms lack cell walls, and all are prokaryotic (lacking a true cell nucleus)

B. Identification
1. Intermediate organisms can be distinguished by serologic tests for nonspecific and specific antigens
2. Although light microscopy using common stains does not usually reveal intermediate organisms, it does occasionally show changes in host cells that suggest infection
3. Standard culture techniques are not always useful for identification because of the slow growth and fastidious nature of the organisms

II. Forms Lacking Cell Walls

A. General information
1. These single organisms have no cell walls
2. They represent resistant phase in the presence of antibiotics, such as penicillin and the cephalosporins, whose mode of action is against cell walls
   a. These antibiotics are usually effective in preventing the growth of cell walls and thereby eliminating most pathogenic microorganisms
   b. To eliminate such microorganisms, researchers must use stronger and potentially toxic antibiotics that attack a different cellular component
3. L-form organisms may revert to typical bacterial forms with cell walls when cultured repeatedly in media without antibiotics or in hosts no longer receiving antibiotic therapy

B. Means of identification
1. Forms lacking cell walls can be identified by careful observation of enriched media for tiny colonies of the organisms
2. When grown in antimicrobial-free media, L-forms can be identified by their reversion to the parent species with cell walls
3. Serologic tests for nonspecific and specific antigens, such as cold agglutinating antibodies and red cell adhesion, also can be used to identify these organisms
   a. In the cold agglutinating antibodies test, patients with primary atypical pneumonia produce antibodies that can agglutinate red blood cells at refrigerated temperatures
   b. In the red cell adhesion test, red blood cells of sheep adhere to the surface of Mycoplasma colonies

C. Location
1. L-forms are commonly found in humans treated with penicillin or cephalosporin antibiotics
2. Mycoplasma, the smallest known free-living organisms, are seen in many animals and humans
3. Cell-wall-deficient organisms have been isolated from humans with infections of the respiratory tract, urinary tract, or blood

D. Importance
1. Forms lacking cell walls include human pathogens and opportunists
2. These organisms are well suited to survive in the presence of antimicrobial agents that are active against cell walls
   a. In some cases, the existence of forms lacking cell walls may explain a reinfection after an apparently successful treatment
   b. Successful treatment of these infections requires the use of antibiotics, such as erythromycin or tetracycline

E. Taxonomic classification: L-forms
1. These organisms demonstrate aerobic to facultative respiration
2. L-forms grow slowly (1 or more weeks) on enriched media and produce pinpoint to microscopic colonies

3. They are derived from many different species of bacteria
    a. The first L-forms discovered were from the parent species *Streptobacillus moniliformis*
    b. Other bacteria produce L-forms when treated with cell-wall-active antibiotics

F. Taxonomic classification: Family Mycoplasmataceae
1. Members of this family exhibit aerobic respiration
2. They occur as single and filamentous forms
3. As with L-forms, mycoplasmas grow slowly (1 or more weeks) on enriched media and produce pinpoint to microscopic colonies
4. An example of these organisms is *Mycoplasma pneumoniae* (the cause of primary atypical pneumonia)

III. Obligate Intracellular Parasites

A. General information
1. Obligate intracellular parasites occur as single coccoid forms
2. As parasites, these organisms require host cells to survive

B. Means of identification
1. Infections typically caused by obligate intracellular parasites are Rocky Mountain spotted fever, urethritis, and conjunctivitis, which produce such symptoms as fever, rash, and local *inflammation*
2. Serologic tests for nonspecific and specific antigens can be used to identify these organisms
    a. Antigens of obligate intracellular parasites can be detected directly by reactions with specific antibodies produced in humans or animals
    b. They can also be detected indirectly by measuring the antibody response in the host
3. Other means of identification are cell cultures, followed by Giemsa and fluorescent stains or electron microscopy

C. Location
1. Obligate intracellular parasites are widespread in nature
2. They can be isolated from infections of the respiratory tract, genitourinary tract, eyes, and blood

D. Importance
1. Obligate intracellular parasites include the human and animal pathogens *Rickettsia* and *Chlamydia*
    a. Chlamydia, a sexually transmitted disease that infects up to 50% of females in the United States, is a symptomless yet severe condition that causes infertility and congenital problems in infants
    b. Rickettsia causes an increasing number of severe and sometimes fatal infections, such as typhus and Rocky Mountain spotted fever

2. *Rickettsia* and *Chlamydia* resemble gram-negative bacteria in structure
   a. The study of these organisms provides insight into the development, metabolism, and infectivity of bacteria
   b. The organisms are highly susceptible to adverse environmental changes and do not survive for long outside a host

E. **Taxonomic classification: Family Chlamydiaceae**
   1. Organisms in this family reproduce only intracellularly from ***elementary bodies***
      a. Elementary bodies are an inactive or dormant yet infectious stage
      b. These bodies develop into active ***reticulate bodies***
   2. Chlamydiae lack the ability to synthesize adenosine triphosphate (ATP), the nucleotide that stores energy in high-energy phosphate bonds
      a. They must acquire ATP from a host in an energy-dependent relationship
      b. All energy for the organism's metabolism and reproduction comes from the host
   3. These microorganisms infect birds and humans
   4. *Chlamydia trachomatis* (the cause of trachoma and increasing numbers of sexually transmitted diseases) belongs to this family

F. **Taxonomic classification: Family Rickettsiaceae**
   1. Rickettsiae, named for the American pathologist Howard Taylor Ricketts, reproduce only intracellularly
   2. Cell cultures are necessary for microbiologic study of these organisms
      a. Because of the difficulty and danger of infection in culturing rickettsia, only the most sophisticated laboratories perform this procedure
      b. All energy for the organism's metabolism and reproduction comes from the host cell's ATP
   3. These organisms are carried by ***vectors*** (ticks, fleas, lice, and mites)
   4. They infect rodents and other small mammals, as well as humans
   5. A member of this family is the tick-borne *Rickettsia rickettsii* (the cause of Rocky Mountain spotted fever)

Study Activities

1. Write a brief paper outlining the ways in which intermediate microorganisms differ from viruses and bacteria.
2. Describe why standard culture techniques are not always helpful in identifying intermediate organisms.
3. Compare L-forms with the bacterium *Streptobacillus*.
4. Compare *Chlamydia* and *Rickettsia* with gram-negative bacilli.
5. Research and report on the increasing incidence of sexually transmitted diseases that result from *Chlamydia*.
6. Find information on the incidence of Rocky Mountain spotted fever in your state. Report your findings to the class.

8

Viruses

Objectives

After studying this chapter, the reader should be able to:
- Define bacteriophage, capsid, oncogenic, and prion.
- Distinguish between viruses and bacteria.
- List the means of identifying DNA viruses, RNA viruses, and prions.
- Describe the characteristics of at least two types of DNA viruses and at least three types of RNA viruses.

I. Virology

A. General information
    1. **Virology** is the study of viruses and virus-like proteins, which are perhaps the smallest living things known
        a. Viruses range in size from 20 to 250 nanometers (nm)
        b. Ordinarily too small to be seen by light microscopy, viruses usually can be detected with electron microscopy
    2. As obligate intracellular parasites, viruses lack the capacity for independent metabolism and reproduce only in living hosts
        a. The hosts that they infect are bacteria, plant cells, animal cells, human cells, and any of the preceding cells in tissue cultures
        b. Viral infection of human cells can cause such diseases as acquired immunodeficiency syndrome (AIDS), chicken pox, hepatitis, and polio
    3. Virus microorganisms have a unique composition
        a. Some viruses are composed of only one type of nucleic acid, which makes up their genetic code; others are composed entirely of protein and contain no nucleic acid
        b. Deoxyribonucleic acid (DNA) viruses are composed of DNA
        c. Ribonucleic acid (RNA) viruses are composed of RNA
        d. Prions are large virus molecules composed of protein organized in filaments or fibrils (small filaments)

B. Viral morphology
    1. The morphology of viruses is simpler than that of bacteria (see *Viral Composition*, page 56)
        a. The virus particle has a rod-shaped central core composed of nucleic acid
        b. Covering the core is a protein coat called a **capsid**

Viral Composition

This schematic drawing shows the main components of a typical polyhedral-shaped virus particle, or virion.

Capsid

Nucleic acid core

Envelope

Spike

---

    c. An envelope derived from the host cell membrane may protect the viral capsids by camouflaging them with host antigens

        (1) Viruses with envelopes tend to be resistant to the host's immune system

        (2) Viruses without envelopes are more common in plants

  2. Individual virus particles are called virions

    a. Virions can be shaped as cubes, spheres, icosahedra (20-sided polygons), or helices

    b. They can also occur as complex forms composed of multiple copies of the shapes mentioned above

C. Viral hosts

  1. Viruses attach to specific *receptor sites* on host cells

    a. The ability of a virus to infect an organism is determined by the types and presence of receptor sites on host cells

    b. Plant viruses tend to infect only plants and animal viruses tend to infect only animals because of the nature of the receptor sites

    c. Receptor sites are proteins on the surface of cells

  2. Once a virus enters a host cell through the receptor site, it takes over the host cell's metabolism and uses the host cell's components to produce more virus particles

    a. DNA viruses are produced directly; the host cell builds viral components from the genetic instructions provided by the viral DNA

    b. RNA viruses induce production of messenger RNA or transcriptase enzymes, which copy viral RNA back into DNA for use by the host cell

  3. The viruses produced by the host cell are then released from this cell, spreading the active infection; in animals and humans, the release of the virus particles prompts the immune system to release interferon and produce antibodies

4. **Incubation periods** for viral diseases range from days to years
    a. Many infections, such as mumps and chicken pox, incubate for a few days, whereas the human immunodeficiency virus may not produce symptoms for 5 to 10 years
    b. Latent, or long-term, infections are possible
        (1) These infections are characterized by asymptomatic periods, followed by the reappearance of symptoms
        (2) Reactivation of a latent virus results in new viral reproduction

II. DNA Viruses

A. General information

1. DNA viruses, which have only deoxyribonucleic acid in their core, typically occur as icosahedral forms
2. After inserting their DNA into the host DNA, the viruses are duplicated during replication, translation, and transcription (see Chapter 4 for more information on DNA replication, translation, and transcription)
3. Cell cultures allow microbiologists to grow these viruses for study
    a. Research with DNA viruses has led to advances in genetics, gene therapy, vaccination and prevention of disease, and cancer treatment
    b. Vaccines to combat hepatitis B, for instance, are produced by recombinant-gene technology
4. DNA viruses often form cellular masses known as **inclusion bodies,** which are helpful in identification

B. Means of identification

1. DNA viruses can be identified by the signs and symptoms of the diseases that they cause
2. If the virus forms inclusion bodies in host cells, these inclusions may be seen when the organism is examined under a microscope
3. Another means of identification are the specific antibodies produced in response to viral antigens; these antibodies can be measured in the blood

C. Importance

1. Active infections caused by DNA viruses include hepatitis, infectious mononucleosis, Burkitt's lymphoma, chicken pox, and smallpox
2. Latent infections caused by DNA viruses include genital herpes and pharyngitis caused by adenoviruses
3. **Oncogenic** (cancer-causing) and potentially oncogenic DNA viruses are common; such viruses cause cancer of the liver and genitourinary tract, as well as lymphoma and papilloma

D. Taxonomic classification: Adenovirus

1. This group of viruses infects mammals and birds through airborne and fecal-oral transmission
2. Adenoviruses affect the upper respiratory tract, intestinal tract, and conjunctivae

   3. At least 100 types of adenoviruses have been identified, five of which cause
      diseases such as pneumonia, conjunctivitis, and gastrointestinal illness in
      children and young adults

E. Taxonomic classification: Bacteriophage
   1. Bacteriophages are viruses that infect bacteria
   2. These viruses attach to the bacteria's cell wall and inject their DNA into the
      bacterial cell
   3. The instructions contained in the viral DNA cause the bacterial cell to produce
      virus particles
   4. Viral replication within the bacterial cell results in *lysis* of the bacteria and
      release of virus particles
      a. The released virus particles are capable of repeating the cycle in other
         bacteria
      b. Rarely, the virus particles are released through the cell membrane and
         wall, leaving the bacteria alive and capable of producing bacteriophages
         indefinitely

F. Taxonomic classification: Herpesvirus
   1. Herpesvirus is transmitted by infected individuals
   2. This group of DNA viruses affects the skin, mucous membranes, and nervous
      system
   3. There are many different types of herpesvirus
      a. Herpes simplex virus (HSV) type I causes fever blisters and cold sores;
         HSV type II causes genital herpes
      b. Cytomegalovirus is a significant pathogen of newborns, organ transplant
         patients, and patients with AIDS; it causes fever, pneumonia, and
         inflammation of internal organs
      c. Epstein-Barr virus causes Burkitt's lymphoma and infectious
         mononucleosis
      d. Varicella-zoster virus causes chicken pox and shingles

G. Taxonomic classification: Poxvirus
   1. Poxviruses infect insects, birds, and mammals and are transmitted by infected
      individuals
   2. They cause rashes and local inflammation of the skin
   3. This group of viruses causes smallpox and cowpox (vaccinia)

H. Taxonomic classification: Papovavirus
   1. Papovaviruses (the name is derived from papilloma, polyoma, and vacuolating
      agents) infect rodents, cattle, primates, and humans
   2. The means of transmission is through infected individuals
   3. These viruses cause hyperplastic lesions and tumors of the skin
   4. Papillomavirus (the cause of epithelial tumors and warts) belongs to this group

I. Taxonomic classification: Hepatitis B virus
   1. The hepatitis B virus infects humans (and possibly woodchucks, ground
      squirrels, and kangaroos) through contact with contaminated blood and other
      body fluids

2. This virus affects liver cells, causing fever, malaise, jaundice, and (occasionally) death
3. It is unrelated to any other known human virus

III. RNA Viruses

A. General information
1. RNA viruses, which have only ribonucleic acid in their core, occur as cubic, icosahedral, *helical,* and spherical forms
2. They contain transcriptase enzymes that copy viral RNA back into DNA; the DNA can then be used for viral replication
3. These viruses often form cellular inclusions that are useful in identification

B. Means of identification
1. RNA viruses can be identified by the signs and symptoms of the diseases that they cause
2. Host cell inclusion bodies may be formed and can be seen during microscopic examination
3. Electron microscopy also is useful for identifying RNA viruses
4. Specific antibodies produced to viral antigens can be measured in the blood
5. Cell cultures permit growth of these viruses for study

C. Importance
1. Active infection caused by RNA viruses includes various diseases of plants, animals, and humans; human diseases include colds, encephalitis, mumps, and measles
2. Latent infections are also possible
3. Oncogenic or potentially oncogenic viruses are common; RNA viruses produce many types of leukemia

D. Taxonomic classification: Arenavirus
1. This class of viruses infects rodents, primates, and humans through airborne transmission and vector bites
2. Arenaviruses affect the central nervous system, blood, lungs, and kidneys
3. Examples include lymphocytic choriomeningitis virus, Lassa fever virus, and South American hemorrhagic fever virus

E. Taxonomic classification: Bunyavirus
1. These viruses infect mammals and humans through bites from mosquitoes and flies
2. Bunyaviruses affect the central nervous system and cause febrile illnesses
3. California virus (a cause of encephalitis) is a bunyavirus

F. Taxonomic classification: Coronavirus
1. These viruses are so named because, under an electron microscope, they resemble a corona or crown
2. Mammals and humans can be infected by airborne transmission of the virus particles

3. Coronaviruses affect the upper respiratory tract in adults and the lower respiratory tract in children
4. Viral strains that cause the common cold belong to this group

G. Taxonomic classification: Influenzavirus
1. The influenzaviruses infect birds, mammals, and humans through airborne transmission
2. This virus causes an acute infection of the respiratory tract
3. Three major strains of influenzavirus have been identified—types A, B, and C

H. Taxonomic classification: Mosaic viruses
1. Mosaic viruses infect plants through airborne transmission or via vectors
2. The virus infection causes hyperplastic lesions on the stems, leaves, and fruit of the plant
3. Tobacco mosaic virus, the cause of tobacco mosaic disease, results in significant economic loss to farmers
4. Because these viruses infect only plants, they provide microbiologists with a safe means of studying viral structure and reproduction

I. Taxonomic classification: Paramyxovirus
1. Paramyxovirus, a subgroup of the myxoviruses, infects birds and humans through airborne and person-to-person transmission
2. The virus particles affect the salivary glands, gonads, skin and mucous membranes, lungs, lymph nodes, central nervous system, and internal organs
3. Paramxyoviruses cause mumps, measles, and rubeola; respiratory syncytial virus (a chief cause of infection in infants); and Newcastle disease

J. Taxonomic classification: Picornavirus
1. These viruses infect many animals, as well as humans, by airborne and oral transmission
2. They affect the respiratory tract, intestinal tract, and nervous system and cause generalized disease
3. Picornaviruses comprise the enteroviruses and the rhinoviruses
   a. Rhinoviruses are the primary cause of the common cold
   b. Other picornaviruses include coxsackieviruses A and B (the cause of fever, meningitis, encephalitis, and paralysis); echoviruses (the cause of acute gastrointestinal disease); and poliovirus types 1, 2, and 3 (the cause of polio)

K. Taxonomic classification: Reovirus
1. Reoviruses, formerly classified as a subgroup of the echoviruses, infect mammals and humans by oral transmission or through tick bites
2. The intestinal tract and central nervous system are affected by these viruses
3. Reoviruses include rotavirus (which accounts for up to 50% of acute gastroenteritis in children younger than age 2) and bivirus (the cause of Colorado tick fever)

L. Taxonomic classification: Retrovirus
1. Retroviruses infect animals and humans through transmission of blood and other body fluids

2. These viruses, which affect white blood cells, the nervous system, the lymphatic system, and internal organs, cause decreased immunity to other infections and cancer
3. Included in this group of viruses are the human T-lymphotropic viruses I and II (the cause of human leukemias) and the human immunodeficiency virus (the cause of AIDS)

M. Taxonomic classification: Rhabdovirus

1. Animals and humans can be infected by these viruses through the bites of infected individuals
2. Rhabdoviruses affect the central nervous system and cause paralysis and convulsions
3. Examples of rhabdoviruses are the rabies virus (the cause of rabies in animals and humans), the Marburg virus (the cause of Marburg disease), and the Ebola virus (the cause of severe hemorrhagic fever)

N. Taxonomic classification: Togavirus

1. Togaviruses are so named because they have envelopes, or "togas"
2. This subgroup of the arboviruses infects animals and humans via transmission from individuals or mosquitoes
3. These viruses affect the skin, mucous membranes, nervous system, and internal organs, causing inflammation, fever, rash, or encephalitis
4. Togaviruses comprise the German measles, or rubella, virus and numerous viruses carried from animals to humans by mosquitoes and ticks

IV. Viruses of Uncertain Designation

A. General information

1. A virus with an unknown nucleic acid composition is classified as "of uncertain designation"
2. Viruses in this category occur as spherical forms 23 to 34 nm in diameter
3. Viruses of uncertain designation are highly resistant to heat, acid, and other chemical treatments

B. Means of identification

1. Electron microscopy can be used to identify these viruses
2. Serologic tests for the presence of specific antigens and antibodies are another means of identification
3. Cell culture techniques are not yet available for viruses of uncertain designation

C. Importance

1. Active infections cause acute gastroenteritis
2. Latent infections and carrier states are possible, which facilitates the spread of infection
   a. A carrier is an individual who has disease-causing microorganisms that can be spread to others
   b. The carrier does not exhibit disease symptoms

D. Taxonomic classification: Norwalk viruses
1. These viruses infect humans through oral transmission
2. Norwalk viruses affect the intestinal tract and cause fever and diarrhea
3. They are responsible for outbreaks of acute gastroenteritis, especially in older children and adults

V. Protein Infectious Particles

A. General information
1. Protein infectious particles, or prions, are glycoproteins with a molecular weight of approximately 30,000
   a. They are the smallest known disease-producing, infectious agents
   b. They are extremely resistant to heat, ultraviolet light, and chemical treatments that kill other viruses
2. Filamentous forms or fibrils of the protein may be formed
3. Prions were first recognized in sheep that had the disease scrapie

B. Means of identification
1. Prions can be identified by the diseases that they cause, such as kuru and scrapie
2. Electron microscopy of infected cells is useful for detecting the presence of prions
3. Prion protein masses can be stained and detected by light microscopy

C. Importance
1. Acute infection by prions is almost always fatal
2. Latent infections caused by prions have incubation periods of several months to many years

D. Taxonomic classification: Prions
1. Prions infect animals and humans through transmission from infected individuals
2. The virus particles affect the central nervous system, causing slowly progressive disease that results in trembling, peripheral nerve disorders, dementia, and death
3. Prions cause scrapie in sheep and hamsters and cause kuru, Jakob-Creutzfeldt disease, and Gerstmann-Straussler-Schwinker disease in humans

Study Activities

1. Describe the basic structure of a typical virus particle.
2. Name five human diseases caused by viruses, and explain how these diseases are diagnosed.
3. Describe the differences between DNA and RNA viruses.
4. List the steps involved in the infection of a host cell by a virus.
5. Research the link between viruses and human cancers.
6. Interview a health professional about the incidence of HIV and AIDS in your state.

9

Fungi

Objectives

After studying this chapter, the reader should be able to:
• Define monomorphic, dimorphic, mycelium, and thallospore.
• Discuss the characteristics that distinguish yeasts and molds from bacteria.
• Describe fungal reproduction.
• List and describe at least one example of monomorphic yeasts, monomorphic molds, dimorphic fungi, and fungi of uncertain designation.

I. Mycology

A. General information

1. *Mycology* is the study of fungi, of which there are three types—yeasts, molds, and mushrooms
2. Mycologists classify fungi as eukaryotic, heterotrophic organisms that lack chlorophyll
   a. More than 100,000 species of fungi have been identified
   b. Only about 100 species are pathogenic to animals and humans
3. Fungi differ from bacteria in several ways
   a. They have a definite nucleus
   b. They occur as single-cell, multicell, or branching forms
   c. They grow in acid pH media, such as Sabouraud dextrose agar
4. Soil and decaying vegetable matter are the main sources of fungi
5. Fungi demonstrate aerobic and facultative respiration

B. Types of fungi

1. Yeasts and yeast-like fungi are single, oval cells and short-filamentous microorganisms that range in size from 5 to 50 micrometers
2. Molds are long, branching, filamentous forms that give rise to macroscopic colonies
3. Larger fungi include mushrooms and puffballs
4. *Monomorphic* (one form) fungi only grow as yeasts or molds
5. *Dimorphic* (two forms) fungi grow as yeasts at 37° C and grow as molds at room temperature

Morphology of Fungi

This schematic drawing shows the major structural elements of typical fungi.

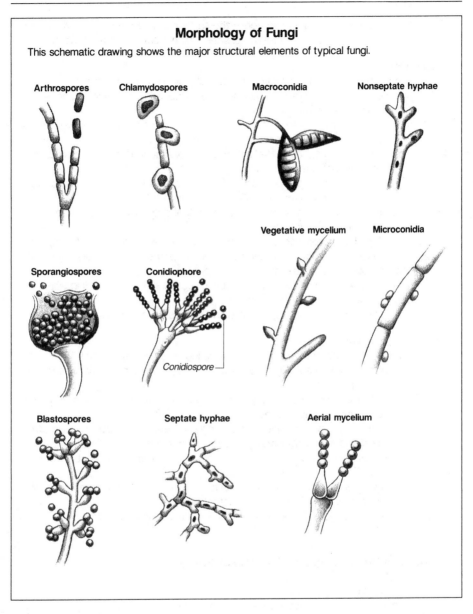

Arthrospores   **Chlamydospores**   **Macroconidia**   **Nonseptate hyphae**

Vegetative mycelium   **Microconidia**

Sporangiospores   **Conidiophore**

Conidiospore

Blastospores   **Septate hyphae**   **Aerial mycelium**

C. Specialized structures

1. A fungus is composed of single tubular filaments called *hyphae* (see *Morphology of Fungi*)
   a. Septate hyphae are divided into individual cells by cross walls
   b. Nonseptate hyphae are undivided
2. All the hyphae that make up a single fungus are collectively known as a ***mycelium***
   a. A vegetative mycelium absorbs nutrients from the environment

b. An aerial mycelium rises above the surface of any medium on which the fungus is growing (for example, the "fuzzy" part of a bread mold)

D. Reproduction

1. Fungi reproduce both sexually and asexually by means of spores
   a. Sexual spores, called *ascospores*, are produced in a saclike structure called the ascus
      (1) Ascospores are produced from the union and fusion of two cells
      (2) Four to eight ascospores may develop in each ascus
   b. Asexual spores, called **thallospores,** are produced directly from one cell in the hyphae (thallus) of the fungus
      (1) Thallospores produced by fragmentation of the hyphae into barrel-shaped forms are called *arthrospores*
      (2) Thallospores produced by swelling of a hyphal cell are called *chlamydospores*
      (3) Thallospores that bud from vegetative cells are called *blastospores*
2. In some fungi, spores are produced in club-shaped structures called *sporangia;* these spores are known as *sporangiospores*
3. Spores produced on stalklike structures, or *conidiophores,* are called *conidia*
4. Each fungal spore can survive adverse conditions such as drying; when conditions improve, they germinate and produce a new fungus

E. Identifying characteristics

1. Mycologists identify fungi by their various individual characteristics
   a. The time it takes a fungus to produce a typical colony is one identifying characteristic
   b. The fungus's odor and color, especially as spores are produced in mature colonies, also help to identify it
   c. Each fungus has a unique morphology on Gram stain, **cotton blue stain, periodic acid-Schiff stain,** capsule stain, or methenamine silver stain
   d. The type of spore formation and biochemical reactions to carbohydrates can be used to identify a fungus
   e. Mycologists also use serologic tests to identify specific antigens on the fungus or on specimens from infected individuals
2. Two broad categories used by mycologists are yeasts and molds

II. Monomorphic Yeasts and Yeast-like Fungi

A. General information

1. Monomorphic yeasts occur as single or budding cells
2. Their pseudohyphae (false hyphae with tubelike structures) and hyphae produce arthrospores, blastospores, and chlamydospores
3. Colonies of monomorphic yeasts mature rapidly and are fully developed in 2 to 5 days

B. Means of identification

1. Monomorphic yeasts can be identified by their color, odor, shape, and type of spore formation

2. Also used for identification are biochemical tests, including the fungus's ability to metabolize sugars, and serologic tests for specific antigens
3. A capsule stain can be used to identify *Cryptococcus* (this yeast-like organism usually has a capsule)

C. Location
1. Monomorphic yeasts are found in soil, air, and places of human habitation
2. These organisms are normal flora of the skin, gastrointestinal tract, and genitourinary tract but can be opportunists of animals and humans with decreased resistance to infections
3. They can be isolated from infections of the skin, hair, gastrointestinal tract, genitourinary tract, and central nervous system

D. Importance
1. Pathogenic monomorphic yeasts cause such diseases as thrush and cryptococcosis, an infection of the central nervous system
2. Nonpathogenic monomorphic yeasts are used to leaven bread and ferment alcoholic beverages

E. Taxonomic classification: Genus *Candida*
1. A member of the family Cryptococcaceae, *Candida* organisms produce creamy white colonies that can be mistaken for bacteria
2. These organisms have a distinctive yeasty odor
3. Examples of this genus are *Candida albicans,* which causes thrush and opportunistic infections in those with acquired immunodeficiency syndrome (AIDS), and *C. tropicalis,* which also causes infections in AIDS patients and those with impaired immune systems

F. Taxonomic classification: Genus *Cryptococcus*
1. These encapsulated organisms produce cream to brown, mucus-like colonies
2. One member of this genus is *Cryptococcus neoformans* types A, B, C, and D (the cause of cryptococcosis)

G. Taxonomic classification: Genus *Geotrichum*
1. *Geotrichum* organisms produce white, moist colonies
2. They develop arthrospores that are useful in identification
3. *Geotrichum candidum* causes geotrichosis, an infection of the mouth, lungs, and intestinal tract

H. Taxonomic classification: Genus *Saccharomyces*
1. These microorganisms produce creamy white, moist colonies
2. Each organism produces one to four ascospores per ascus
3. This genus includes *Saccharomyces cerevisiae,* also known as brewers' or bakers' yeast

I. Taxonomic classification: Genus *Trichosporon*
1. Members of this genus produce cream to gray colonies, at first moist and then becoming wrinkled
2. These organisms produce arthrospores and blastospores

3. Although *Trichosporon* organisms are normal flora of the respiratory and gastrointestinal tracts of humans and animals, *Trichosporon cutaneum* causes white piedra, an infection of the hair shafts

III. Monomorphic Molds

A. General information
1. Monomorphic molds produce all types of spores
2. Colonies of monomorphic molds reach maturity in 3 to 21 days
3. These microorganisms include saprophytes, opportunists, and pathogens
   a. Dermatophytes are superficial pathogens
   b. Subcutaneous or systemic pathogens infect the internal organs

B. Means of identification
1. When grown on culture, monomorphic molds can be identified by their color and shape
2. Spore formation is another means of identifying these organisms

C. Location
1. Monomorphic molds are widespread in soil, air, food, and places of human habitation
2. They can be isolated from infections of the skin and internal organs

D. Importance
1. Monomorphic molds are normal flora in plants and on the skin and hair of animals and humans
2. Pathogenic molds can cause a variety of plant, animal, and human infections, especially in individuals with decreased resistance to infections
3. These molds are the source of antibiotics that control other microorganisms (see Chapter 13 for more information)

E. Taxonomic classification: Class Maduromycetes
1. These organisms produce colonies with surfaces that are white to brown or black with a brown or black vegetative mycelium
2. They produce chlamydospores, ascospores, and microconidia
3. This class includes the genera *Monosporium, Phialophora,* and *Madurella* (the cause of maduromycosis, an infection of the skin and internal organs)

F. Taxonomic classification: Class Phycomycetes
1. This group of fungi, which includes the common water, leaf, and bread molds, produce white to gray colonies
2. These molds typically produce nonseptate hyphae with sporangia
3. This class includes the genera *Absidia, Mucor,* and *Rhizopus* (the causes of phycomycosis or zygomycosis, a systemic infection in those with compromised immune systems)

G. Taxonomic classification: Genus *Alternaria*
1. These microorganisms produce colonies that are white to greenish black or brown on the surface with a black vegetative mycelium

2. Their septate hyphae produce chains of septate macroconidia
3. *Alternaria* includes saprophytes and common contaminants of the environment
   a. This organism causes several plant diseases
   b. It is also a common allergen in human bronchial asthma

H. Taxonomic classification: Genus *Aspergillus*
1. The colonies produced by *Aspergillus* are first white then yellow, green, brown, or black with a white to brown vegetative mycelium
2. They have septate hyphae that produce ***sterigmata*** covered with chains of conidia resembling a paint brush
3. Members of this genus include saprophytes and *Aspergillus fumigatus* (the cause of the lung infection aspergillosis)

I. Taxonomic classification: Genus *Microsporum*
1. *Microsporum* produce colonies that are white to yellow-brown with a yellow or reddish vegetative mycelium
2. They form microconidia and septate macroconidia
3. *Microsporum audouinii* (the cause of epidemic ringworm of the scalp in children) and *Microsporum canis* (the cause of skin and nail infections) belong to this group

J. Taxonomic classification: Genus *Penicillium*
1. Colonies of the genus *Penicillium* are white to blue-green with a white vegetative mycelium
2. They have septate hyphae that produce multiple branching conidiophores with chains of conidia
3. *Penicillium notatum* is a source of the antibiotic penicillin

IV. Dimorphic Fungi

A. General information
1. These single and budding organisms grow as yeast-like cells at 37° C and as molds at room temperature (22° to 30° C)
2. Dimorphic fungi form chlamydospores and conidia
3. Colonies mature slowly, requiring 4 to 28 days to become fully developed

B. Means of identification
1. Dimorphic fungi can be identified by the type of growth that they exhibit at 37° C and at room temperature
2. They also can be identified by their color, shape, and type of spore formation
3. Microbiologists can use serologic tests for specific antigens and antibodies; for example, antibodies to *Blastomyces* and *Histoplasma* can be detected in the blood of infected individuals

C. Location
1. Dimorphic fungi are widespread in soil, air, and places of human habitation
2. They can be isolated from infections of the skin and respiratory tract
3. *Blastomyces* is found in the soil of North America and Africa

    4. *Histoplasma* is most common in the soil of river valleys and caves throughout
      the world

D. Importance
    1. Dimorphic fungi include human pathogens that cause diseases of the skin and
      internal organs
    2. Spores produced in the mold phase disseminate the fungi and spread
      infections

E. Taxonomic classification: Genus *Blastomyces*
    1. These dimorphic fungi produce white to brown colonies
    2. They have single microconidia on septate hyphae
    3. *Blastomyces dermatitidis* (the cause of blastomycosis, an infection of the skin,
      lungs, and bone) belongs to this genus

F. Taxonomic classification: Genus *Coccidioides*
    1. The colonies produced by *Coccidioides* are white to brown
    2. Because they produce highly infectious arthrospores, culturing these organisms
      is hazardous
    3. Coccidioidomycosis, a respiratory or systemic infection prevalent in the
      southwestern United States, is caused by *Coccidioides immitis*

G. Taxonomic classification: Genus *Histoplasma*
    1. These organisms produce white to brown colonies
    2. They have microconidia and knobby macroconidia on septate hyphae
    3. This genus includes *Histoplasma capsulatum* (the cause of the African form of
      histoplasmosis, a pulmonary infection)

H. Taxonomic classification: Genus *Sporothrix*
    1. These dimorphic fungi produce white to brown or pink colonies
    2. They have conidia that appear in a flowerette arrangement on septate hyphae
    3. One member of this genus is *Sporothrix schenkii* (the cause of sporotrichosis,
      an infection of the skin and lymph glands)

V. Fungi of Uncertain Designation

A. General information
    1. The characteristics of some fungi make identification and classification difficult
    2. Additional study often results in the proper classification of these
      microorganisms; until then, they are known as fungi of uncertain designation
    3. Included in this group are delicate branching forms of fungi that may separate
      on cell division into bacteria-like forms
      a. Separate forms rarely develop in infections
      b. Similar microorganisms include *Nocardia*

B. Means of identification
    1. Fungi of uncertain designation are identified by their culture characteristics,
      such as color and shape

2. Another means of identification is the examination of stained smears for hyphae breaking up into segments that resemble bacteria
3. Microbiologists also use biochemical tests to identify these organisms

C. Location
1. Fungi of uncertain designation are distributed worldwide in soil
2. They can infect animals and humans and can be isolated from infections of the skin

D. Importance
1. These fungi cause animal and human disease
2. Infections of the skin and systemic infections in immunocompromised individuals are common

E. Taxonomic classification: Genus *Actinomadura*
1. The colonies produced by *Actinomadura* are white, tan, pink, or red
2. These organisms were formerly classified as members of the bacteria genus *Streptomyces*
   a. A better understanding of the basic differences between bacteria and fungi allows microbiologists to clarify classifications
   b. Continued study—and name changes—can be expected in this area
3. *Actinomadura* cause mycetoma, a tumor-like infection of the feet with multiple draining abscesses

Study Activities

1. Name two of the relatively few pathogenic fungi, and describe their characteristics.
2. Identify three single-cell fungi, three multicell fungi, and one branching fungi.
3. Explain how thallospores develop.
4. With your instructor, examine cultures of fungi. Describe your findings.
5. Make microscopic smears of fungi and stain them. Draw examples of your observations.
6. Research the discovery of the antibiotic penicillin from the mold *Penicillium*.

10

Protozoa and Protozoa-like Organisms

Objectives

After studying this chapter, the reader should be able to:
• Define cilia, cyst, protist, pseudopodia, and trophozoite.
• Distinguish between protozoa and other microorganisms.
• Describe at least two classes of amebae, flagellates, ciliates, sporozoans, and algae.

I. Protozoology

A. General information

1. Members of the kingdom Protista, protozoa and protozoa-like organisms are large, single-cell microorganisms
   a. Before being categorized as **protists,** protozoa were classified as single-cell plants and animals
   b. They range is size from 10 micrometers to several millimeters (mm)
2. These eukaryotic, heterotrophic or photosynthetic organisms demonstrate aerobic respiration
3. Many protozoa are free-living, while others live in mutual or parasitic relationships with plants, insects, animals, and humans
4. Protozoa are found in fresh and salt water, soil, plants, insects, animals, and humans
5. They produce a variety of infections of plants, animals, and humans
   a. Plant infections caused by protozoa include heartrot of coffee and cocoanut
   b. Human infections caused by protozoa include diarrhea, dysentery, malaria, pneumonia, and toxoplasmosis
6. Protozoa and protozoa-like organisms are classified according to their method of locomotion
   a. Amebae move by means of pseudopodia (a protrusion of cytoplasm from the ameboid cell)
   b. Flagellates move by means of flagella (fine, threadlike structures that protrude from the cell)
   c. Ciliates move by means of cilia (short, hairlike structures that protrude from the cell)
   d. Sporozoans are nonmotile protozoa
   e. Protozoa-like algae move by means of flagella

B. Growth and reproduction

1. Protists usually reproduce asexually
   a. A ***trophozoite*** is the active, usually motile, feeding stage of an immature protozoon
   b. Trophozoites reproduce asexually by fission
2. Active trophozoites may form a resistant, but infectious, resting stage known as a *cyst*
   a. Encysted protozoa are nonmotile
   b. They can survive adverse conditions, such as drying
   c. Up to eight new trophozoites may be produced within each cyst
3. Some protozoa reproduce by sexual means
   a. Sexual reproduction can occur by simple conjugation
   b. It also can occur through the production of specialized male and female ***gametocytes,*** the fusion of which results in a new protozoan
   c. Sexual reproduction occurs in both free-living and parasitic forms

II. Class Rhizopoda (Amebae)

A. General information

1. These single-cell organisms have both free-living and parasitic forms
2. They move by means of pseudopodia, protrusions of cell membrane and protoplasm by which the organism pulls itself along
3. Amebae reproduce by binary fission

B. Means of identification

1. Microbiologists can examine amebae in fresh wet preparations or stained smears
   a. Amebae have a distinct size and morphology
   b. On microscopic examination, nuclear ***chromatin*** can be seen in distinctive patterns
2. When examining cysts of intestinal parasites, a microbiologist can identify up to four (*Entamoeba histolytica*) or up to eight (*Entamoeba coli*) nuclei in each cell

C. Location

1. Amebae are found throughout the world
2. They can be isolated from infections of the intestinal tract and internal organs

D. Importance

1. Amebae are the simplest single-cell organisms
2. They include normal human flora, as well as pathogens that cause gastrointestinal disease and abscesses of the liver, lung, and brain
3. They also include organisms called foraminifers
   a. Foraminifers have chalky, many-chambered shells
   b. Accumulations of these shells are responsible for major geologic formations, such as limestone beds and the White Cliffs of Dover, England

Amebae

The schematic drawing below depicts a typical ameba and cysts in the class Rhizopoda.

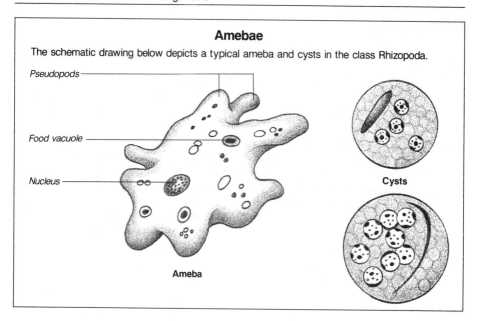

Pseudopods

Food vacuole

Nucleus

Ameba

Cysts

E. Taxonomic classification: Order Foraminifera
1. These free-living amebae live in salt water
2. They produce a calcium carbonate shell, which is the source of lime
3. Geologic deposits of limestone and chalk are created by the accumulation of these organisms' shells

F. Taxonomic classification: Genus *Amoeba*
1. As free-living organisms, amebae inhabit soil, water, and plants
2. This genus includes the largest ameba, which is 1 to 2 mm in size
3. An example of this group is *Amoeba proteus,* which is used by microbiologists to study *chemotaxis* (movement of an organism in response to chemical agents) and *phagocytosis* (engulfment of particles by *phagocytes*)

G. Taxonomic classification: Genus *Entamoeba*
1. These organisms exist as normal flora and intestinal parasites of animals and humans
2. They are transmitted from host to host by the fecal-oral route
3. This genus includes *Entamoeba coli* (a nonpathogenic commensal and occasionally opportunistic organism) and *Entamoeba histolytica* (the cause of amebic dysentery and abscess of the liver; it can spread to other organs via the bloodstream)

H. Taxonomic classification: Genus *Naegleria*
1. These free-living microorganisms inhabit fresh water
2. They are classified as ameboflagellates
   a. During its life, this organism alternates between an ameboid phase and a phase during which it possesses two flagella

b. It can be found in large numbers in shallow freshwater, especially in warm weather
3. *Naegleria fowleri* is an opportunistic ameboflagellate that infects swimmers during the summer; it causes a usually fatal meningoencephalitis

III. Class Mastigophora (Flagellates)

A. General information
1. Most of these single-cell, oval-shaped organisms are free-living in soil and water
2. Some flagellates live in mutualistic or parasitic relationships
3. They move by means of one or more flagella and reproduce by binary fission

B. Means of identification
1. Flagellates can be identified by their size and morphology
2. They can also be identified by the types of disease that they cause
   a. Severe cases of diarrhea are caused by flagellates
   b. Sleeping sickness is an infection of these microorganisms
3. Animal inoculation may be helpful for identifying *Leishmania* and *Trypanosoma*
   a. Mice and other laboratory animals are injected with the microorganisms
   b. The blood, bone marrow, and tissues of the animals provide smears for study
4. Serologic tests can be used to identify specific antigens
   a. For example, fluorescent antibody staining can be used to identify *Leishmania*
   b. Antibodies to *Trypanosoma* in the blood of patients are useful in diagnosis
5. Skin tests can be used to detect exposure to these microorganisms
   a. In these tests, antigen from a microorganism such as *Leishmania* is injected into the skin
   b. If previous exposure has occurred, antibodies react with the antigen, producing redness and swelling at the injection site

C. Location
1. Flagellates are distributed worldwide
2. They can be found in infections of the skin, blood, intestinal tract, and internal organs

D. Importance
1. Flagellates include normal gastrointestinal flora and organisms that are pathogenic for many animals and humans
2. These microorganisms cause many deaths and major disabilities in underdeveloped countries where poor hygiene, lack of sanitation, and insect vectors spread infection

E. Taxonomic classification: Order Dinoflagellata
1. Dinoflagellates are free-living organisms found in seawater
2. Because they contain chlorophyll, these organisms can produce food from light through photosynthesis
3. These plantlike protozoa also contain red pigments

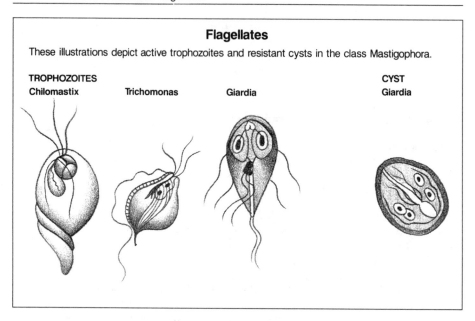

Flagellates

These illustrations depict active trophozoites and resistant cysts in the class Mastigophora.

TROPHOZOITES

Chilomastix   Trichomonas   Giardia

CYST

Giardia

4. This order includes the species *Gonyaulax catenella* and *Gymnodinium breve* (the causes of algal blooms, or "red tide")

F. Taxonomic classification: Genus *Euglena*
1. These free-living flagellate protozoa live in fresh water but are found in abundance in stagnant water
2. They contain chlorophyll, which is lost as the *Euglena* grows under saprophytic conditions
   a. Chlorophyll is the green pigment in plants in which photosynthesis occurs
   b. As a saprophytic, *Euglena* takes in food from its surroundings
3. *Euglena viridis,* which is considered to share the characterisitics of plants and animals, is commonly studied in laboratories
   a. Chlorophyll is a plant characteristic
   b. Lack of a cell wall and motility by flagella are animal characteristics

G. Taxonomic classification: Genus *Giardia*
1. *Giardia* are parasitic protozoa that can be isolated from animal and human intestinal tracts
2. Found as cysts in contaminated water, these organisms are infectious by the fecal-oral route
3. Included in this group is *Giardia lamblia* (the cause of giardiasis, a protracted diarrhea)

H. Taxonomic classification: Genus *Leishmania*
1. These parasitic organisms live in the skin and lymphatic systems of mammals and humans in tropical regions
2. Biting sandflies carry the organisms from host to host

   3. This genus includes *Leishmania tropica*, *L. braziliensis*, and *L. donovani* (the causes of leishmaniasis, an ulcerative disease of the skin, mucous membranes, and internal organs)

I. **Taxonomic classification: Genus *Trichomonas***
   1. These pear-shaped organisms live as commensals in the gastrointestinal tracts and as parasites in the genitourinary tracts of animals and humans
   2. They are transmitted directly from host to host
   3. *Trichomonas hominis* is considered normal flora in oral cavities, but *T. vaginalis* is the cause of sexually transmitted vaginitis

J. **Taxonomic classification: Genus *Trichonympha***
   1. These mutualistic protozoa are found in the intestinal tracts of termites
   2. *Trichonympha* can digest the cellulose in wood
   3. These organisms enable termites to feed on wood

K. **Taxonomic classification: Genus *Trypanosoma***
   1. These protozoan parasites live in the blood, lymph nodes, and central nervous systems of animals and humans
   2. They are transmitted from host to host by insects
   3. Among the members of this genus are several human pathogens
      a. *Trypanosoma rhodesiense* and *T. gambiense*, which are transmitted by the tsetse fly, cause African sleeping sickness
      b. *T. cruzi*, which is carried by reduviid bugs, causes Chagas's disease, a form of trypanosomiasis

IV. Class Ciliata (Ciliates)

A. **General information**
   1. Members of the Class Ciliata are the most highly specialized and complicated of the protozoa
      a. These single-cell organisms live in colonies and occur as both free-living and parasitic forms
      b. They are characterized by their hairlike structures, known as cilia, which provide motility
   2. Ciliates are unique in that they possess two nuclei
      a. The macronucleus controls cell metabolism
      b. The micronucleus controls cell division
   3. Ciliates also possess a mouth or gullet for ingesting food, another unique feature of this class
   4. These organisms reproduce asexually (by fission) and sexually (through conjugation)

B. **Means of identification**
   1. Microbiologists can identify ciliates by their characteristic size and morphology
   2. These organisms can also be identified by their means of motility; the rows of moving cilia are a unique feature

Ciliates

These illustrations depict active trophozoites and resistant cysts in the class Ciliata.

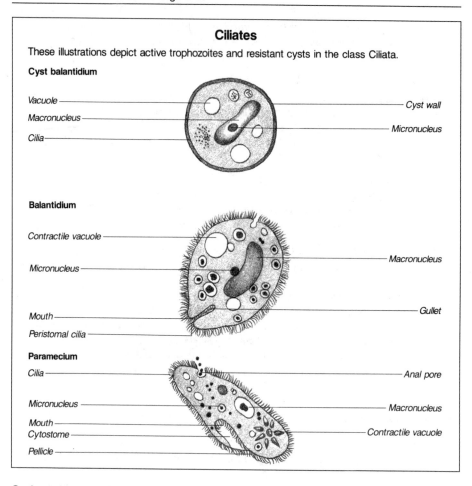

Cyst balantidium

Vacuole — — Cyst wall

Macronucleus —

— Micronucleus

Cilia —

Balantidium

Contractile vacuole —

Micronucleus — — Macronucleus

Mouth — — Gullet

Peristomal cilia —

Paramecium

Cilia — — Anal pore

Micronucleus — — Macronucleus

Mouth —
Cytostome — — Contractile vacuole

Pellicle —

C. Location
1. Ciliates are found throughout the world
2. They can be isolated from infections of the intestinal tract

D. Importance
1. The Class Ciliata includes human pathogens that cause gastrointestinal disease
2. Ciliates also provide microbiologists with the opportunity to study simple sexual reproduction by conjugation

E. Taxonomic classification: Genus *Balantidium*
1. These organisms are parasites of the human intestinal tract
2. *Balantidium* organisms are transmitted via the fecal-oral route
3. One member of this genus, *Balantidium coli,* is the largest protozoan parasite of humans
   a. It is 50 to 100 micrometers long and 40 to 70 micrometers wide
   b. This species is a cause of dysentery

F. **Taxonomic classification: Genus** *Paramecium*
1. These free-living protozoa can be found in fresh water
2. They feed on bacteria, smaller microorganisms, and other particulate matter
3. Members of the genus *Paramecium* are used by microbiologists to study motility, phagocytosis, and conjugation

V. Class Sporozoa (Sporozoans)

A. **General information**
1. Sporozoa are single-cell, obligate organisms
   a. All sporozoa are parasites
   b. adults, they are characterized by their lack of cilia or flagella
2. A two-host life cycle is common, as illustrated by *Plasmodium,* which causes malaria
   a. *Plasmodium* organisms are transferred from one host (a female mosquito) to another host (a human) through insect bites
   b. *Plasmodium* organisms can live and multiply within a mosquito's body without harming the host
   c. Once transferred to a human, they first enter the liver cells and then the red blood cells, where they multiply repeatedly
   d. A mosquito that bites an infected human can begin the cycle again
3. Sporozoans reproduce asexually (by fission) and sexually (through gametocytes)

B. **Means of identification**
1. Thin and thick smears of blood from people infected with malaria are used to identify these parasites
2. Histologic examination of lung biopsies are used to identify sporozoa in people infected with pneumonia
3. Serologic tests can be used to identify specific antigens, such as antibodies to toxoplasmosis

C. **Location**
1. Sporozoa are distributed worldwide
2. They can be found in infections of the blood, respiratory tract, and internal organs

D. **Importance**
1. Sporozoa are a significant cause of disease in animals and humans
   a. Malaria, caused by *Plasmodium,* is a leading cause of death and disability in underdeveloped countries
   b. Pneumocystis pneumonia, caused by *Pneumocystis,* is often seen in patients with acquired immunodeficiency syndrome (AIDS)
   c. Toxoplasmosis, caused by *Toxoplasma,* is responsible for neonatal diseases and death
2. These microorganisms, all parasites of animal hosts, cannot ingest food other than by simple **diffusion**

Sporozoa

The illustrations below depict infectious trophozoites and cysts in the class Sporozoa.

E. Taxonomic classification: Genus *Plasmodium*
   1. These organisms lead a two-host life cycle and have different means of reproduction, depending on the host
      a. The plasmodium reproduces sexually within a female mosquito host (male mosquitos do not require blood to survive and therefore do not transmit the organisms)
         (1) This sexual stage, or sporogony, begins when male and female gametocytes fuse within the mosquito
         (2) Sporogony results in fertile cysts that release infectious sporozoites into the saliva of the mosquito
         (3) A mosquito bite passes sporozoites into the bloodstream of a new host
      b. Within a human or animal host, the plasmodium reproduces asexually
         (1) This asexual stage, or schizogony, takes place within the host's cells
         (2) Each trophozoite can produce 8 to 24 infectious merozoites in red blood cells and up to 40,000 in liver cells
   2. Members of this genus include *Plasmodium falciparum, P. malariae, P. ovale,* and *P. vivax,* all of which cause malaria

F. Taxonomic classification: Genus *Pneumocystis*
   1. Although currently classified in the class Sporozoa, *Pneumocystis* may be more closely related to the fungi than to the sporozoa
   2. These organisms often colonize in patients who have impaired immune systems

3. *Pneumocystis* organisms produce round, infectious forms in the lungs of infected individuals
   a. These infectious particles are spread by airborne transmission
   b. Latent infections are so common that virtually all children have been exposed to *Pneumocystis* by age 4
4. *Pneumocystis carinii* (the cause of severe pneumonia in immunosuppressed hosts, especially AIDS patients) belongs to this genus

G. Taxonomic classification: Genus *Toxoplasma*

1. *Toxoplasma* organisms are parasites of children and immunosuppressed patients
2. They produce intracellular inclusions in birds, animals, and humans; cats and cattle are major sources of infection
3. *Toxoplasma gondii,* a member of this genus, causes toxoplasmosis, a febrile illness involving the internal organs, especially those of newborns

VI. Phylum Chlorophyta (Protozoa-like Algae)

A. General information

1. These single-cell, colonial organisms occur as free-living forms that live in salt water
2. As photosynthetic organisms, algae contain chlorophyll
3. These organisms move by means of flagella
4. As with other protozoa, algae reproduce asexually (by fission) and sexually (through gametocytes)
5. Unlike other protozoa, algae possess cell walls

B. Means of identification

1. Algae can be identified by their distinctive size and morphology
2. The presence of chlorophyll within the cell is another identifying characteristic

C. Location

1. Algae can be found throughout the world
2. They exist in fresh and salt water, in soil, and on moist surfaces

D. Importance

1. The phylum Chlorophyta is important because it may represent progressive steps in the development of more complex and multicellular organisms
   a. The characteristics exhibited by Chlorophyta lie between those of protozoa and those of higher life forms, thus making these organisms useful for study
   b. To investigate sexual reproduction, microbiologists can study *Volvox,* which reproduces via eggs and sperm
2. Lichens, which turn rocks into soil, result from a mutualistic relationship involving algae and fungi

E. Taxonomic classification: Genus *Chlamydomonas*

1. These organisms occur as single, flagellated cells

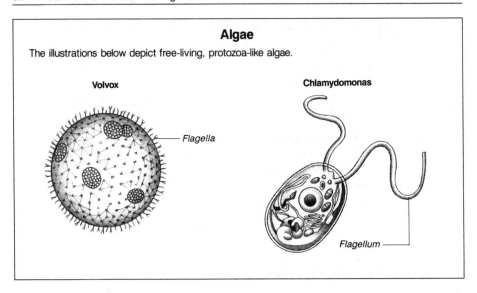

Algae

The illustrations below depict free-living, protozoa-like algae.

Volvox

Flagella

Chlamydomonas

Flagellum

2. They have a red eyespot that seeks out light for photosynthesis
3. This group includes microorganisms that are typical of the most primitive protists; these organisms are useful for the study of organism development

F. **Taxonomic classification: Genus *Volvox***
1. Although they live in spherical colonies, individual *Volvox* cells resemble *Chlamydomonas* cells
2. They reproduce sexually through the production of sperm and eggs (gametocytes)
3. This group includes microorganisms that have characteristics more developed than those of protists and less developed than those of multicellular forms

Study Activities

1. Lead a class discussion on the development of six diseases caused by parasitic protozoa.
2. Describe the characteristics of algae that protozoa do not possess.
3. With your instructor, grow a hay infusion and examine it for protozoa.
4. Describe the types of motility seen in protozoa.
5. Research and report on the impact of malaria on human history.
6. Interview a health professional on protozoan diseases in your area. Report your findings to the class.

11

Helminths

Objectives

After studying this chapter, the reader should be able to:
• Define helminth, parthenogenesis, larvae, and copepod.
• Name and describe the three types of helminths.
• Distinguish between helminths and other microorganisms.
• List and describe at least one genus of rotifers, roundworms, tapeworms, and flukes.

I. Helminthology

A. General information
1. *Helminths* are worms and wormlike, multicellular organisms
2. Some are microscopic, whereas others are many meters in length
3. They have organ systems for digestion, circulation, locomotion, sensation, and reproduction
   a. Reproduction occurs asexually (fission, budding, and *parthenogenesis*) and sexually (egg and sperm)
   b. The eggs and immature stages, or *larvae,* are microscopic
4. Helminths can be found in fresh and salt water, soil and decaying vegetation, plants, animals, and humans
5. They include free-living and parasitic organisms
   a. Free-living helminths have a highly developed morphology but a simple life cycle
   b. Parasitic helminths have a simplified morphology but a complex life cycle that often involves two or more hosts
6. These organisms cause a variety of diseases, especially anemia and weight loss from intestinal infections

B. Types of helminths
1. Helminths are categorized on the basis of their body structure
   a. Aschelminthes (rotifers) have a saclike body structure
   b. Nemathelminthes (nematodes or roundworms) have a threadlike body structure
   c. Phatyhelminthes have a flat, segmented, or tapelike body structure (cestodes or tapeworms) or a flat, unsegmented, and leaf-shaped body structure (trematodes or flukes)

2. Helminths range in size from the microscopic to the largest parasites of humans, up to 10 meters in length

II. Class Rotifera (Aschelminthes)

A. **General information**
   1. Rotifers, or aschelminths, are elongated sacks of protoplasm that contain multiple nuclei
   2. Cilia around the oral cavity aid in feeding and locomotion and, when vibrating, resemble rotating wheels
   3. Mature rotifers sometimes build tubelike shells
   4. Although rotifers have separate sexes, reproduction can occur asexually by parthenogenesis

B. **Means of identification**
   1. Rotifers can be identified by their body structure, which is transparent and up to 100 micrometers long
   2. The organism's internal structures are readily apparent when it is examined under the microscope
   3. These helminths are characterized by cilia that form one or more circles around the organism's head

C. **Location**
   1. Rotifers inhabit stagnant, fresh water
   2. They are distributed worldwide

D. **Importance**
   1. Rotifers, the simplest helminths, represent a possible transitional step between protists and multicelled animals
   2. Microbiologists study the rotifer's sensory and reproductive systems
      a. These systems represent the simplest development of specialized tissues in multicell organisms
      b. The entire, living organism can be examined with the aid of a microscope

E. **Taxonomic classification: Genus *Rotifer***
   1. These free-living organisms live in fresh, still water
   2. They feed on microscopic plants and animals
   3. Rotifers can survive many years in a dried condition while waiting for water
   4. One member of this genus, *Rotifer vulgaris,* is a commonly studied microorganism that inhabits pools and ponds

III. Class Nematoda (Roundworms)

A. **General information**
   1. More than 3,000 species of roundworms, or nematodes, have been identified
      a. A shovel-full of garden soil may contain up to 1 million roundworms
      b. Adult roundworms range in size from 1.5 mm to 35 cm

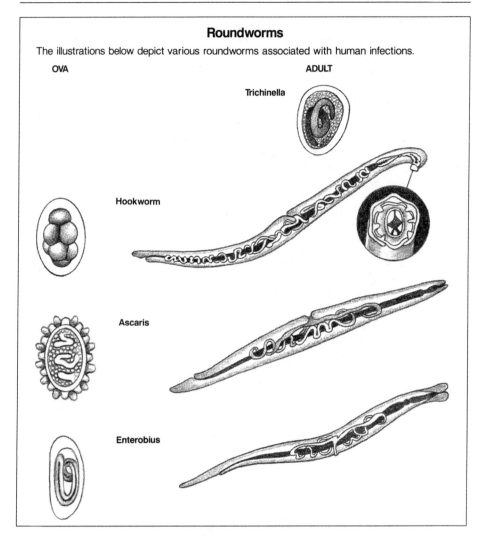

Roundworms

The illustrations below depict various roundworms associated with human infections.

OVA   ADULT

Trichinella

Hookworm

Ascaris

Enterobius

  c. Female roundworms are about twice as large as male roundworms
 2. Roundworms are so named because they are round in cross section
 3. Both free-living and parasitic forms exist
 4. Roundworms reproduce sexually; the female typically lays thousands of
   microscopic eggs, or ova

B. Means of identification
 1. A roundworm's eggs can be identified in the feces of infected hosts
 2. Pinworms can be identified by collecting their eggs on cellophane tape
   touched to the perianal skin of the host
 3. Filariae and *Trichinella* are identified through blood smears and tissue biopsies
 4. Serologic tests based on specific antigens of *Toxocara* can detect antibodies in
   the blood of infected individuals

C. Location
1. Roundworms are distributed worldwide but are most prevalent in tropical and subtropical areas
2. They are common in underdeveloped countries with poor hygiene, sanitation, and education

D. Importance
1. Roundworms produce a variety of animal and human diseases, including heartworm and hookworm infection, elephantiasis, and blindness
2. They are a leading cause of death and disability in underdeveloped countries
3. Each year, roundworms cause billions of dollars worth of damage to crops and livestock in the United States

E. Taxonomic classification: Genus *Ascaris*
1. These parasites live in the human intestinal tract
   a. Adults lay infective eggs
   b. Eggs are transmitted by the fecal-oral route
2. *Ascaris* organisms are the largest human intestinal roundworms; adults grow to 35 cm in length
3. *Ascaris lumbricoides,* a roundworm that resembles the earthworm, is the cause of ascariasis
   a. Ascariasis produces colicky pains and diarrhea, especially in children
   b. This illness affects approximately 1 billion people worldwide and approximately 4 million people in the United States

F. Taxonomic classification: Genus *Enterobius*
1. *Enterobius* organisms are parasitic roundworms that colonize the lower intestinal tracts of humans
2. These organisms cause nighttime itching and insomnia
   a. These symptoms are caused by the female worm, which migrates to the host's anus to lay eggs on the perianal skin
   b. Eggs shed from the skin into the environment are infectious
3. This genus includes *Enterobius vermicularis,* also known as the pinworm
   a. Up to 200 million people worldwide and nearly 40 million people in the United States are infected by pinworms
   b. Pinworm infection is common in children

G. Taxonomic classification: Filaria
1. Filaria is a loosely applied generic name for certain parasites of the blood, lymphatic systems, and tissues of animals and humans
2. Filaria are transmitted from infected hosts to new hosts by mosquitoes and biting flies
3. These organisms include *Brugia malayi* and *Wuchereria bancrofti* (the cause of elephantiasis), *Onchocerca volvulus* (the cause of river blindness), *Loa loa* (the source of eyeworm infection), and *Dirofilaria immitis* (the cause of heartworm infections in dogs and cats)

H. Taxonomic classification: Hookworms
1. These parasitic organisms live in the human intestinal tract

2. Humans can become infected by the hookworm larvae that hatch from eggs in contaminated soil
   a. The larvae penetrate the skin of the feet and migrate to the intestinal tract
   b. Proper sanitation and wearing of shoes can prevent most hookworm infections
3. Hookworms include *Necator americanus,* or New World hookworm, and *Ancylostoma duodenale,* or Old World hookworm

I.   Taxonomic classification: Genus *Strongyloides*
   1. *Strongyloides* are free-living in soil and parasitic in human intestinal tracts
   2. Their eggs hatch in the host's intestines
      a. Larvae are passed in the feces, and those that survive in contaminated soil are infectious for new hosts
      b. The larvae can penetrate human skin on contact
   3. These organisms reproduce asexually by parthenogenesis
   4. *Strongyloides stercoralis,* the cause of strongyloidiasis (which affects 400,000 people in the United States alone), belongs to this group

J.   Taxonomic classification: Genus *Toxocara*
   1. *Toxocara* organisms contaminate the environment with infective eggs
      a. If the eggs are ingested by a dog or cat, this roundworm can become a parasite in the animal's intestines
      b. Because humans are inappropriate hosts, *Toxocara* organisms are unable to complete their life cycle within the human body
   2. Members of this genus include *Toxocara canis,* which is commonly found in dogs, and *T. cati,* which is commonly found in cats

K. Taxonomic classification: Genus *Trichinella*
   1. These parasitic roundworms can be found as infective cysts within the muscles of pigs, bears, rats, and humans
   2. *Trichinella* is transmitted when infected meat is eaten raw or has not been properly cooked
      a. Within the host's intestines, the cyst dissolves and a larva emerges
      b. When the parasite matures, it migrates to the muscles
   3. *Trichinella spiralis,* also called pork worm, belongs to this genus
      a. At 1.5 mm in length, it is one of the smallest roundworms
      b. *T. spiralis* is the cause of trichinosis

L. Taxonomic classification: Genus *Trichuris*
   1. *Trichuris* organisms are parasites of the human intestinal tract
   2. Their infective eggs develop in the soil before being ingested by new hosts
   3. *Trichuris trichiura,* also known as the whipworm, is the species that principally infects humans
      a. This roundworm is approximately 2 inches in length
      b. It can cause diarrhea and vomiting in infected hosts

M. Taxonomic classification: Genus *Turbatrix*
   1. These free-living organisms are found in fruits, vegetables, juices, beer, and vinegar

2. Adult females are less then 2 mm in length and give birth to live larvae only 25 micrometers long
3. Included in this genus is *Turbatrix aceti,* a small roundworm found in vinegar

IV. Class Cestoda (Tapeworms)

A. General information
1. Tapeworms are segmented, flat worms
   a. There are approximately 10,000 species of flatworms, two-thirds of which are parasitic
   b. They range from 6 cm to many meters in length
2. Because tapeworms lack a mouth or digestive system of their own, they absorb nutrients through their body walls from the digested food of the host
3. Tapeworms have a head, or scolex, and body segments known as proglottids
   a. The head of the tapeworm has suckers and hooks, which are used to attach to the intestinal tract of the host
   b. At maturity, proglottids contain ovaries and testes and can act as independent reproductive entities

B. Means of identification
1. Tapeworms can be identified by the presence of microscopic eggs in the host's feces
2. Occasionally, mature proglottids are found in the feces

C. Location
1. Tapeworms can be found throughout the world
2. They commonly thrive in areas of inadequate hygiene, sanitation, and education and in areas where fish and water plants are eaten raw or undercooked

D. Importance
1. Tapeworms cause disease in animals and humans
   a. They can cause illness by encroaching on the host's food supply
   b. They can also cause problems by producing wastes and obstructing the host's intestinal tract
2. They cause substantial economic loss and disability in underdeveloped countries

E. Taxonomic classification: Genus *Diphyllobothrium*
1. Members of this genus inhabit fresh water, where their eggs hatch
   a. The *Diphyllobothrium* larvae then infect intermediate hosts—**copepods** (tiny crustaceans) and fish
   b. They are transmitted to their final hosts—dogs, cats, minks, bears, and humans—when infected fish is eaten raw or undercooked
2. *Diphyllobothrium latum,* the fish tapeworm, belongs to this group
   a. This species can be ¾ inch wide and grow up to 30 feet (about 10 meters) long

Tapeworms

These illustrations depict structural characteristics of two typical tapeworms. Although the adult forms have a different scolex, their ova are virtually identical.

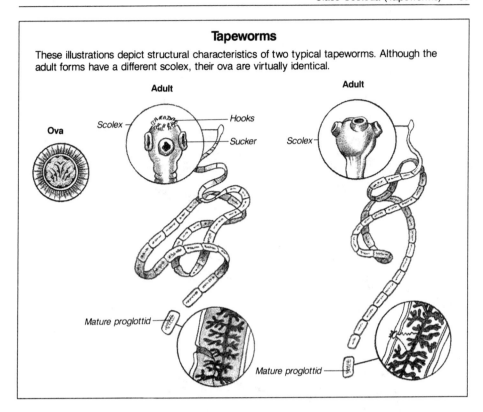

b. Humans infected by the fish tapeworm exhibit symptoms similar to those of pernicious anemia

F. Taxonomic classification: Genus *Hymenolepis*
1. These parasites can infect the intestinal tracts of rodents and humans
2. They range in size from 20 to 80 mm
3. They produce eggs that are infectious when ingested by a suitable host
4. *Hymenolepis nana,* the dwarf tapeworm, is a member of this genus

G. Taxonomic classification: Genus *Taenia*
1. *Taenia* are large tapeworms that live as parasites in the muscles and intestinal tracts of swine, cattle, and humans
   a. Their infective cysts are transmitted in raw or undercooked meat
   b. When the meat is eaten, the larvae are released and develop in the intestinal tract
2. This genus includes *Taenia saginata,* the most common large tapeworm in humans, and *T. solium,* the pork tapeworm, which is rarely found in the United States

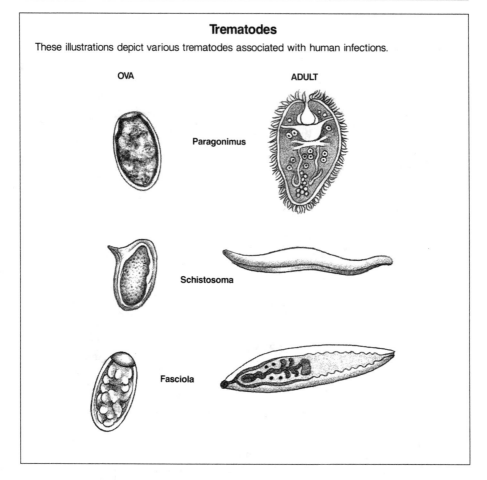

Trematodes

These illustrations depict various trematodes associated with human infections.

OVA ADULT

Paragonimus

Schistosoma

Fasciola

V. Class Trematoda (Flukes)

A. General information
1. These leaf-shaped, unsegmented worms range in size from 5 to 30 mm
2. Parasitic flukes attach to a host by means of suckers
    a. Unlike tapeworms, flukes have an oral sucker through which they draw nourishment from the host
    b. A second sucker secures the fluke to a surface
3. Flukes reproduce by sexual means

B. Means of identification
1. The fluke's microscopic eggs can be found in the feces, sputum, or urine of infected hosts
2. Adult parasitic flukes are occasionally found in the blood and internal organs
3. Free-living forms can be identified in freshwater samples

C. Location
1. Flukes are distributed worldwide
2. They are found in standing freshwater

D. Importance
1. Flukes cause a variety of animal and human diseases, including schistosomiasis and infections of the lungs, liver, and circulatory system
2. They are responsible for considerable death and disability in underdeveloped countries
3. The planaria are used for biologic research in sensation, learning, and regeneration

E. Taxonomic classification: Genus *Clonorchis*
1. Members of the genus *Clonorchis* hatch in freshwater
   a. They pass through two intermediate hosts – snails and fish – before infecting their final hosts – dogs, cats, and humans
   b. Within the final hosts, these flukes develop in the intestinal tract and migrate to the gallbladder and liver
2. *Clonorchis sinensis,* the Chinese liver fluke, belongs to this group

F. Taxonomic classification: Genus *Fasciola*
1. These parasitic organisms hatch in freshwater and pass through an intermediate host, a snail, before colonizing the internal organs of cattle, sheep, and humans
   a. *Fasciola* organisms are transmitted to the final host through the ingestion of infected water plants
   b. Eggs produced within the final host's internal organs enter the intestinal tract
   c. The larvae develop within the intestinal tract, then migrate to the gallbladder and liver
2. The giant liver fluke, *Fasciola hepatica,* is a member of this genus
   a. This fluke is found in sheep, oxen, goats, and horses
   b. When found in the human liver, it can obstruct the bile ducts

G. Taxonomic classification: Genus *Paragonimus*
1. Flukes in this genus hatch in freshwater and initially inhabit intermediate hosts, first a snail, then a crab or crayfish
2. When mammals and humans ingest an infected intermediate host, the parasite remains in the final host's internal organs
   a. Eggs produced within the internal organs enter the gastrointestinal tract
   b. After developing in the gastrointestinal tract, the organisms migrate to the lungs
3. This genus includes *Paragonimus westermani,* the lung fluke
   a. This oval or pear-shaped fluke causes parasitic hemoptysis
   b. Infection by this fluke is especially prevalent in Asiatic countries

H. Taxonomic classification: Genus *Planaria*
1. These free-living flukes live in freshwater

2. Because of their particular characteristics, members of this genus are often used in studies of simple nervous systems, learning, and regeneration
   a. The planarian has two eyespots on its head that react to light, making it useful in the study of sensation
   b. A planarian trained to react to certain stimuli (such as avoiding electric shock) can be cut up and fed to an untrained planarian, which is then able to perform as though trained
   c. A planarium cut in half will regenerate the missing part

I. **Taxonomic classification: Genus *Schistosoma***
   1. *Schistosoma* organisms hatch in freshwater but require snails as intermediate hosts
   2. The free-swimming larvae penetrate the skin of the final host — water birds or humans — and enter the host's circulatory system
      a. The eggs produced within the host's vascular system and internal organs enter the intestinal or urinary tracts
      b. Eggs passed into freshwater hatch and begin the life cycle anew
   3. Included in this genus are microorganisms that cause swimmer's itch and schistosomiasis (the single most serious helminth infection, affecting about 200 million people worldwide)

Study Activities

1. Name the four taxonomic classes of helminths.
2. Distinguish between nematodes and cestodes.
3. Examine specimens of helminths provided by your instructor. Describe your observations.
4. Research and report on hookworm infections.
5. Research and report on tapeworm infections.
6. Interview a health professional on helminthic diseases in your area.

12

Infection and Epidemiology

Objectives

After studying this chapter, the reader should be able to:
• Define virulence and epidemiology.
• Distinguish between infectivity and virulence.
• List and describe the four host defenses used to resist infection.
• Name five ways in which diseases are transmitted.

I. Resistance

A. **General information**
   1. *Infection,* the introduction of parasitic microorganisms into a host, results in injury or death to the host cells
   2. Infected host cells can be injured in several ways
      a. The microorganisms may compete with the host cell for nutrients, as in tapeworm infections
      b. The microorganisms may release **toxins,** as in *Clostridium tetani* infections
      c. The microorganisms may subvert host cell processes, as in viral infections
      d. The microorganisms may trigger antigen-antibody reactions, which can result in cell death and inflammation, as in syphilis and gonorrhea
   3. *Resistance* is the ability of an organism to withstand infection
      a. Higher life forms, including humans, are relatively resistant to infection
      b. Factors involved in resistance include the host's general state of health, personal hygiene, nutrition, psychological factors such as stress, and sanitation conditions in the environment
   4. Specific defenses used to ward off infection include physical defenses, cellular defenses, antibodies, and normal flora

B. **Physical defenses against infection**
   1. Physical barriers prevent disease-producing microorganisms from entering sites where infection can occur
   2. Intact skin, tears, mucus, and the acidity of the stomach and genitourinary tract are examples of physical barriers
   3. Physical defenses can be compromised by cuts and abrasions to the skin, burns, trauma, and excessively wet or dry conditions in the environment

C. Cellular defenses against infection

1. Resistance to infection is enhanced by specific cells that can search out and destroy pathogenic microorganisms
2. Neutrophils, monocytes, and macrophages are cells that can phagocytize and lyse microorganisms
3. Infection can occur when there are too few phagocytic cells or when cell function is ineffective in producing lysis

D. Antibody defenses against infection

1. Antibodies are proteins that are produced in response to and react with antigens
   a. They are made by lymphocytes that come in contact with a microorganism's antigens
   b. Antibodies and antibody-like proteins protect higher life forms from pathogenic microorganisms
2. An antibody tends to react only with a specific antigen; exceptions are called cross-reactions
3. Antigen-antibody reactions result in agglutination (clumping) of the microorganisms, neutralization or destruction of the microorganisms, and increased phagocytosis
4. Antibody-like proteins that are nonspecific in their reactions include interferon (active against viruses), lysozyme (active against some gram-positive bacteria), and properdin (active against gram-negative bacteria and viruses)
5. Antibody defenses can fail when antibody production is deficient or genetically impaired

E. Normal flora as a defense against infection

1. Normal flora are harmless microorganisms that reside on the skin and in the gastrointestinal and genitourinary tracts
   a. They are found in many of the sites where infectious agents can cause disease
   b. Normal flora compete with pathogenic microorganisms for nutrients
2. The presence of normal flora stimulates the immune system before infections occur
   a. During and after birth, the newborn acquires organisms of the normal flora
   b. The normal flora stimulate development of the reticuloendothelial system (responsible for fighting infections) and increase antibody levels
3. The indiscriminate use of antimicrobial agents can eradicate normal flora, thereby eliminating this defense against infection

II. Infectivity

A. General information

1. Infectivity is the ability of a microorganism to overcome resistance and establish a parasitic relationship with a host
2. Infection occurs only when specific conditions are met
   a. Of primary importance is the microorganism's ability to survive on or in host tissues

   b. Nutrients and suitable environmental conditions are also required
   c. An ability to grow at body temperatures is fundamental (dermatophyte fungi cannot invade deeper tissues because of their inability to grow at 37° C)
   d. Adherence to cell surfaces is also advantageous; the pili of bacteria, the receptor sites of viruses, and the hooks or suckers of helminths allow such adherence

B. Pathogenic nature of microorganisms
   1. More than 100 years ago, bacteriologist Robert Koch recognized that culturing colonies of a single organism was vital to the study of microorganisms
   2. He developed a system of related ideas, known as Koch's Postulates, that still serve to prove the pathogenic nature of microorganisms
      a. The same microorganism must be found in all cases of a disease
      b. The microorganism must be isolated in pure culture from the infected host
      c. The microorganism must reproduce the disease when introduced into a susceptible host
      d. The microorganism must be isolated in pure culture from the experimentally infected host

C. Special characteristics and infectivity
   1. Some organisms have special characterisitics that contribute to their infectivity
      a. The presence of a capsule helps to deter phagocytosis and antibody formation; *Cryptococcus, Klebsiella,* and *Streptococcus* are encapsulated organisms
      b. The presence of mucolytic and proteolytic enzymes can neutralize host defenses; *Haemophilus* possesses such a characteristic
      c. The ability to alter surface antigens and evade antibody reactions can increase infectivity; *Borrelia* and *Neisseria* have this ability
      d. The ability to mask surface antigens so that they resemble host antigens is another way of increasing infectivity; *Escherichia coli and Treponema* have this characteristic
      e. Some pathogenic organisms can overcome the strict conservation of iron that is typical of host cells, freeing the iron for the metabolic needs of the organisms
   2. More than one of these characteristics may be combined in a single, highly infectious organism

III. Virulence

A. General information
   1. Virulence refers to the disease-producing, or pathogenic, ability of microorganisms
      a. An organism's virulence depends on the degree of cell damage and inflammation produced
      b. Virulence also is a function of the release of toxins that directly affect host cell processes

   2. Virulence can be increased through latent infections, which is the case with many viruses

   3. In some instances, an infection with one type of microorganism enhances the virulence of other microorganisms; for example, infection with the human immunodeficiency virus allows opportunists to cause life-threatening diseases in the infected host

B. Production of disease

   1. Direct cell damage is caused by obligate intracellular parasites, such as *Chlamydia, Rickettsia,* and viruses

     a. Cell damage results in acute inflammation

     b. Examples include streptococcal pneumonia and viral hepatitis

   2. The production of toxins, such as the **endotoxins** of gram-negative bacteria and the **exotoxins** of gram-positive bacteria, results in sepsis

   3. The production of certain enzymes, such as the coagulase produced by *Staphylococcus,* can contribute to an organism's pathogenicity

C. Opportunism

   1. Opportunists are considered to be of low virulence

     a. They are commonly members of the normal flora

     b. They produce infection only when introduced into a normally sterile body site, when populations of flora are upset, or in immunocompromised hosts

   2. Opportunism increases with increasingly latent infection rates

IV. Epidemiology

A. General information

   1. Epidemiologists use the tools of microbiology to study the causes of outbreaks, or epidemics, of infectious diseases

   2. They determine how infections are transmitted and recommend preventive measures, such as quarantine, vaccination, or improved sanitation and education

   3. Epidemiologists are employed by health care institutions, local and state health departments, and federal agencies such as the Public Health Service and the Centers for Disease Control

B. Sources of infection

   1. Community-acquired infections are the most common

     a. These infections commonly are self-limiting, require no treatment, and pose no threat to life

     b. Examples include colds and such childhood diseases as mumps and measles

   2. Hospital-acquired, or nosocomial, infections are spread in a health care setting

     a. Many of these infections are life-threatening and difficult to treat

     b. Examples include staphylococcal infections in newborns and burn patients and *Pseudomonas* infections in patients with respiratory disease

C. Disease transmission

1. Diseases are transmitted from an infectious agent to a susceptible host in several ways
   a. Inanimate objects (fomites), such as eating utensils, hypodermic needles, or contaminated clothes, can harbor pathogenic organisms
   b. Vectors, such as fleas, lice, mites, ticks, flies, and mosquitoes, can be a source of disease transmission
      (1) The vector acquires the pathogenic parasite from an animal, referred to as a reservoir
      (2) The vector then passes the parasite to a new host
   c. Human carriers can incubate an infection or remain symptom-free while transmitting an infection to others
   d. Contaminated food or water can spread disease
   e. Airborne agents, such as bacteria, viruses, fungal spores, and helminth eggs, can transmit disease
   f. Direct contact with a dead or diseased organism can be a cause of infection
2. The chain of infection comprises the relationships between fomites, vectors, reservoirs, carriers, and hosts

D. Preventing infection

1. Quarantining infected persons can limit the spread of disease
2. Vaccinating susceptible hosts can stimulate antibody formation
3. Disposing of contaminated wastes properly can discourage the transmission of disease
4. Educating people in the value of personal hygiene, the causes of infection, and the means of controlling exposure to pathogenic microorganisms can help control disease transmission

Study Activities

1. Describe five physical defenses against infection.
2. Name the three cells that act as phagocytes. Describe phagocytosis.
3. Research and report on antibody production in response to an infection.
4. Compare and contrast fomites and vectors.
5. Interview a school nurse on the problems of lice and common childhood diseases.

13

Control of Microorganisms

Objectives

After studying this chapter, the reader should be able to:
- Define antiseptic, disinfection, immunity, pasteurization, and sterilization.
- Discuss how microbistatic control measures differ from microbicidal control measures.
- Give at least three examples of microbial control by physical, chemical, and medical means.

I. Mechanisms of Control

A. General information
1. Microorganisms must be controlled to prevent infection, treat diseases, limit spoilage and decay, and regulate useful processes such as fermentation
2. *Microbistatic* control measures inhibit the reproduction and growth of microorganisms
3. *Microbicidal* control measures kill the microorganisms

B. Microbistatic control
1. A microorganism's reproduction and growth can be inhibited by depriving it of water, nutrients, or a suitable temperature
2. Drying, refrigeration, and freezing are microbistatic control measures

C. Microbicidal control
1. Microorganisms can be killed by disrupting their metabolism, structural integrity, or means of reproduction
2. Bacteria, viruses, and fungi are killed by bactericides, viricides, and fungicides, respectively
3. *Disinfection* kills only harmful organisms, while *sterilization* kills all organisms
4. Boiling, incineration, and use of chemicals are microbicidal control measures

D. Implications of microbial control
1. Because microorganisms develop stages resistant to heat, drying, and adverse conditions, effective control measures must overcome these survival strategies
   a. Spores produced by bacteria and fungi, cysts produced by protozoa, and thick-shelled eggs produced by helminths are examples of resistant stages

　　　b. Steam and gas sterilization are two methods of eliminating resistant
　　　　　stages
　2. Control techniques should effect the desired result without destroying more
　　　than is necessary and without creating adverse environmental consequences

II. Control by Physical Means

A. General information
　1. Physical control measures are effective against a wide variety of
　　　microorganisms
　2. These techniques can be microbistatic or microbicidal
　3. Physical means include removal, moist heat, desiccation, incineration, low
　　　temperatures, radiation, and sanitation

B. Direct removal
　1. Direct removal of a microorganism physically eliminates it from potential sites of
　　　growth
　2. *Washing* with soap and water is perhaps the single most important measure
　　　for controlling and removing microorganisms
　3. *Flushing* carries microorganisms away; for example, tears are effective in
　　　flushing microorganisms from the eyes
　4. *Filtration,* which depends on the size of openings in the filters, removes
　　　microorganisms too large to pass through
　　　a. The smallest microorganisms, such as viruses, are filterable; that is, they
　　　　　can pass through filters
　　　b. Bacteria and similar microorganisms are effectively removed by filtration

C. Moist heat
　1. Moist heat is a microbicidal technique that involves the use of high
　　　temperatures and humidity to effect protein **denaturation**
　2. The effectiveness of moist heat depends on the temperature, duration of
　　　exposure, and pressure used
　3. *Steam sterilization* is usually done in an autoclave, a device that maintains
　　　steam at a temperature of 121° C under 15 pounds of pressure
　　　a. Items must be sterilized for at least 15 minutes to ensure that all
　　　　　microorganisms have been killed
　　　b. Canning, or sealing foods in containers, uses a steam sterilization process
　4. *Pasteurization* is a two-step moist heat treatment used for milk and other foods
　　　that cannot withstand high temperatures
　　　a. The first heating kills actively growing organisms and causes any spores
　　　　　that are present to germinate
　　　b. A second heating kills most of the remaining microorganisms
　　　c. Temperatures range from 63° C for 30 minutes to 72° C for 15 seconds,
　　　　　followed by rapid cooling

D. Desiccation

1. Desiccation, a microbistatic measure, is the removal of water necessary for a microorganism's growth
2. Drying in warm air, *lyophilization*, and freeze drying are effective means of desiccation
3. Manipulating *osmotic pressure*, as with high concentrations of salt or sugar (curing), renders water unavailable for microbial metabolism

E. Incineration

1. Incineration, exposing contaminated materials to flame, is a microbicidal technique
2. It is useful for sterilizing metal instruments, glass microscope slides, and flammable materials, such as paper
3. Incineration has the added advantage of reducing the volume of waste for disposal

F. Low temperatures

1. Low temperatures slow or stop a microorganism's metabolic processes
2. Refrigeration, at temperatures ranging from 1° to 6° C, and freezing, at temperatures of $-18°$ C or less, are microbistatic control measures

G. Radiation

1. Radiation, a microbicidal method, is the emission of energy in waves (light or sound) or particles (ionizing radiation)
2. *Ultraviolet light* directly alters a microorganism's deoxyribonucleic acid
   a. This type of radiation is useful for treating air, surfaces of objects, and thin layers of liquids, such as donated blood
   b. Its usefulness is limited by its low penetrating power; the interior of solid objects and volumes of liquids are not treated
3. *Sonication* is the use of high-frequency sound to disrupt cells
   a. It is used for materials that cannot otherwise be treated
   b. Frequencies up to 40 kilohertz are generated
4. *Ionizing radiation* disrupts a microorganism's metabolism, cellular integrity, and genetic processes
   a. It is used for long-term sterilization of foods
   b. It is also effective for sterilizing pharmaceutical products

H. Sanitation

1. Sanitation is the use of control measures to ensure the public health
2. It involves a combination of physical control techniques
   a. Washing is used for personal hygiene
   b. Filtration is used in water treatment
   c. Drying is used in sewage treatment
   d. Incineration is used for garbage disposal
   e. Radiation is used for decontaminating waste by sunlight, artificial ultraviolet light, or ionizing radiation

III. Control by Chemical Means

A. **General information**
   1. Chemical control measures are effective against all microorganisms
   2. These microbicidal techniques include antiseptics, disinfectants, and sanitizers
      a. *Antiseptics* prevent the growth of any microorganism
      b. *Disinfectants* destroy pathogenic microorganisms
      c. *Sanitizers* reduce the number of microorganisms to safe levels as determined by public health officials
   3. Chemical agents act by disrupting cell membranes, as with surface-active agents, such as alcohols and soaps; or by combining with or destroying proteins, as with alcohols, hydrogen peroxide, *halogens,* metal salts, and acids
   4. The chemical agent's effectiveness is affected by the concentration of the chemical; the length of time the organism is exposed to the chemical; the temperature (chemicals are more potent at higher temperatures); the pH of the surroundings; and the presence of interfering substances, such as high levels of protein

B. **Alcohols**
   1. Alcohols, such as ethanol, propanol, and isopropanol, are liquids that are used as 70% solutions in water
   2. Alcohols reduce surface tension and destroy proteins and the integrity of cell membranes
   3. They are used to disinfect skin

C. **Detergents and soaps**
   1. Detergents and soaps are wetting agents and chemical emulsifiers available in bar, granule, and liquid forms
   2. They are used in water solutions to reduce surface tension and destroy proteins
   3. They are used to disinfect skin, fabrics, glassware, and medical instruments

D. **Ethylene oxide**
   1. Ethylene oxide is a penetrating gas that destroys proteins
   2. It is used to sterilize items that cannot be heated, such as paper and plastics

E. **Aldehydes**
   1. Aldehydes, such as formaldehyde and glutaraldehyde, are 5% to 10% solutions in water
   2. They act by destroying proteins
   3. They are used to treat skin infections and for the long-term preservation of biologic materials

F. **Hydrogen peroxide**
   1. Hydrogen peroxide is a 3% solution in water that acts by oxidizing protein
   2. It is used to disinfect the skin

3. It is less effective than other chemical means because it can be easily neutralized; many bacteria produce enzymes that break down hydrogen peroxide to water and oxygen

G. Halogens

1. Halogens are solutions of chlorine, hypochlorite (bleach), or iodine that oxidize protein
2. They are used to disinfect skin and surfaces and to purify water
3. To be effective, they must be applied for at least 30 seconds

H. Metal salts

1. Metal salts, such as arsenic, copper, mercury, silver, and zinc, are available as ointments, powders, or solutions
2. They act by oxidizing protein and are used to disinfect the skin

I. Phenol

1. Phenol (carbolic acid) and its derivatives (cresols and hexachlorophene) are available in solutions, soaps, and ointments
2. They reduce surface tension and destroy proteins
3. They are used to disinfect skin and surfaces

IV. Control by Medical Means

A. General information

1. Medical means are effective against pathogenic microorganisms
2. These microbistatic and microbicidal techniques include preventing and treating infection, establishing immunity, and administering antibiotics

B. Preventing infection

1. Using physical and chemical controls during hospitalization, surgery, or dental procedures is one means of preventing infection
2. Isolating infected individuals can minimize the spread of disease from person to person
3. Quarantining the sources of infection limits the exposure of susceptible populations to the disease
4. Vaccinations can be used to prepare an individual's immune system to resist infection
5. Educating the public about the causes and spread of disease can also help to prevent infection

C. Establishing immunity

1. Immunity is the state of being resistant to infection or the effects of infectious microorganisms
2. Immunity is determined by antibodies
   a. *Active immunity* occurs when an individual produces his or her own antibodies; it results from exposure to a disease or from vaccination

    b. *Passive immunity* occurs when an individual receives antibodies from another source; antibodies can be transmitted through breast milk or by injection

3. Hypersensitivity or allergy is an excessive response by the body to an antigen-antibody reaction

D. Treating infection

1. When a patient is diagnosed as having an infection, health care workers seek to return the patient to health
2. Physical and chemical means of control are employed to combat the infection
   a. Infected tissue, such as a gangrenous limb or an abscessed tooth, can be physically removed
   b. Antimicrobial therapy can be used to eradicate the infecting organisms

E. Use of antibiotics

1. Antibiotics are natural products of bacteria and molds that inhibit or kill other microorganisms
   a. The first antibiotic, penicillin, was discovered by Alexander Fleming in 1929
   b. Sulfonamides, or sulfa drugs, were discovered in 1933
   c. These early antibiotics were improved by chemical alteration to make them easier to absorb, more soluble, and less toxic to the patient
2. Modern antibiotics are classified by type and mode of action
   a. Penicillins and cephalosporins are active against cell walls
   b. Macrolides, aminoglycosides, and tetracyclines are active against the protein components of microorganisms
   c. Polymyxins and amphotericins are active against cell membranes
   d. Other antibiotics are active against enzymes or compete for vital nutrients such as folic acid
3. Occasionally two antibiotics work better together than alone
   a. This effect, known as synergy, takes advantage of the fact that different antibiotics have different modes of action
   b. Because of synergy, the commonly used antibiotics penicillin and tetracycline may be more effective when administered together than either one alone
4. Indiscriminate or ineffective use of antibiotics can result in microorganisms that are resistant to many antibiotics
   a. Infection with these "super microbes" is difficult if not impossible to treat with antibiotics
   b. Stronger drugs, surgery, and other means may be needed to treat such infections

Study Activities

1. Provide four rationales for controlling microorganisms.
2. Give three examples of microbistatic control measures and microbicidal control measures.
3. Keep a journal of the control measures you use for 1 week.
4. With your instructor, perform an antibiotic susceptibility test. Record your observations.
5. Research and report on hypersensitivity.
6. Interview a health professional on the prevention of an infectious disease, such as AIDS.

Appendices

Appendix A: Glossary

Appendix B: Efficacy of Antibiotics Against
Susceptible Pathogens

Selected References

Index

Appendix A: Glossary

Aerobe—organism that requires oxygen to survive

Agar—solid culture medium

Anaerobe—organism that lives and grows in the absence of oxygen

Antibiotic—microorganism-produced substance that has the capacity to inhibit or kill other microorganisms

Antigen—component of microorganisms and other cells that induces an antibody response in a host

Antimicrobial—substance that limits the growth of microorganisms

Antiseptic—substance that prevents infection by inhibiting the growth of microorganisms without necessarily killing them

Autotroph—organism capable of synthesizing organic molecules from inorganic substances

Bacterial spore—resting stage of a microorganism

Bacteriology—study of bacteria

Binomial—having two names

Broth—liquid culture medium

Capsid—protein coat of a virus that protects its core

Cell—fundamental functional and structural unit of living organisms

Chemotaxis—movement toward a chemical stimulus

Chromatin—pattern of material that makes up a nucleus

Chromosome—threadlike structure in cells that contains genetic information

Cilia—minute, hairlike structures that propel certain microorganisms by vibratory motions

Conjugation—sharing of genetic material between two cells

Copepod—any one of numerous, minute, aquatic crustaceans

Cotton blue stain—mycologic stain composed of lactophenol cotton blue

Darkfield microscopy—passing light around the edge of a specimen being viewed through a microscope

Denaturation—destruction of the usual nature of a substance

Diffusion—spontaneous, random mixing of molecules of different substances

DNA probe—means of identifying organisms based on their genetic composition

Elementary body—dormant but infective phase of *Chlamydia* that develops into a reticulate body

Endotoxin—poisonous substance present in intact cells of gram-negative bacteria

Enzyme—protein that regulates the rate of chemical reactions

Eukaryote—cell that contains a true nucleus

Exotoxin—poisonous substance excreted by bacteria into the surrounding medium

Facultative—having the ability to adjust to various conditions

Fastidious—having complex nutritional requirements

Fermentation—anaerobic reactions in which pyruvic acid is reduced to form such products as ethyl alcohol and lactic acid

Filamentous—resembling fine thread

Fission—means of asexual reproduction in which a cell splits into two daughter cells; also called binary fission

Flagella—fine, threadlike structures whose whiplike motions propel certain microorganisms

Flora—all the plant organisms in a given location; in microbiology, the bacteria usually associated with a site (for instance, the normal flora of the skin)

Fluorescence—property of giving off light, usually in the visible range, when light of a different wavelength is absorbed

Gametocytes—male and female cells required for sexual reproduction

Gene—biologic unit that contains the hereditary characteristics of an organism

Halogen—any of the nonmetallic, highly reactive elements (chlorine, iodine, bromine, fluorine, or astatine)

Helical—having the shape of a spiral or coil

Helminth—wormlike, multicell organism

Hemin—nutrient provided by hemoglobin from red blood cells

Hemolysis—destruction of red blood cells

Heterotroph—organism that cannot manufacture organic compounds and that must rely on other organisms for food

Host—any organism that provides a growth site for another organism

Hyperplasia—abnormal growth in the number of cells in a tissue

Inclusion body—any unusual structure within a cell, formed by cellular material or a parasite

Incubation period—time from infection to onset of symptoms

Inert—not reacting to other elements

Inflammation—tissue response to injury, consisting of pain, heat, redness, swelling, and loss of function

Inoculum—microorganisms introduced into a medium or host

Larva—immature stage of an organism that must change form as it becomes an adult

Lens—piece of glass or other optically clear material shaped to focus or scatter rays of light

Log—exponent denoting the power to which a number is raised; also called logarithm

Lyophilization—rapid freezing and dehydration in a vacuum

Lysis—destruction, as of bacterial cells by a bacteriophage

Magnifying power—ratio of apparent image size (as through a microscope) to actual size

Medium—nutritive substance used to culture microorganisms

Microbiology—study of microorganisms

Microorganism—living organisms too small to be seen by the unaided eye; also called microbes or germs

Microscope—instrument for visualizing organisms that cannot be seen by the unaided eye

Morphology—study of the shape and structure of an organism

Motility—ability to move

Mycelium—mat or mass of fungal hyphae

Mycology—study of yeasts and molds

Nitrogen fixation—process by which atmospheric nitrogen is converted first to ammonia and finally to nitrates

Nomenclature—process of naming organisms

Nucleus—structure in many cells that contains the organism's genes

Obligate—organism that can survive only in a particular environment

Oncogenic—causing tumor formation, especially by viruses

Opportunist—microorganism that does not usually cause disease but under certain conditions can become pathogenic

Organelle—any specialized structure in or on a cell

Osmotic pressure—balance of forces in solutions separated by a semipermeable membrane

Oxidant—electron acceptor in an oxidation reaction

Oxidation—chemical combination with oxygen or removal of hydrogen; loss of an electron by an atom

Parasite—organisms living in or on a host organism

Parasitology—study of parasites

Parfocal—type of light microscope that remains in focus when the objective lenses are changed

Parthenogenesis—reproduction by development of an unfertilized egg

Pathogen—organism that causes disease

Periodic acid-Schiff stain—histologic stain useful for identifying fungi in tissues

pH—hydrogen ion concentration that indicates the degree of acidity or alkalinity of a solution or medium

Phagocyte—cell that ingests microorganisms, other cells, and foreign particles

Phagocytosis — surrounding and engulfing material as typified by feeding protozoa

Phase-contrast microscopy — use of polarized light as a light source in a microscope to enhance contrast and improve visualization

Pili — fingerlike projections used by cells to anchor themselves to surfaces in their environment

Prokaryote — organism, such as a bacterium or blue-green alga, that does not contain a true nucleus

Protist — any single-cell microorganism or member of the kingdom Protista

Receptor site — specific location on a host cell to which a virus particle must be absorbed before infecting the cell

Reduction — chemical combination with hydrogen or removal of oxygen; gain of an electron by an atom

Resolving power — ability to form distinguishable images of objects separated by small distances (as in a microscope)

Reticulate body — active phase of *Chlamydia* that metabolizes and reproduces

Saprophyte — organism that lives on dead or decaying organic matter

Sepsis — condition in which pathogenic microorganisms or their toxins infect the blood or tissues of a host

Smear — small amount of a specimen

Specimen — organism that is representative of its particular species

Spore — cell capable of developing into an adult organism without fusion with another cell; also, a body formed within bacteria and regarded as the resting stage of the cell

Sterigmata — swollen or club-shaped end of a fungal cell from which reproductive elements are produced

Substrate — substance acted on by an enzyme

Symbiosis — close relationship between two dissimilar organisms

Taxonomy — study of the systematic classification of organisms based on their categories, or taxa

Thallospore — reproductive structure that develops directly from the hyphae of fungi

Toxin — poisonous substance capable of inducing antibody formation

Transduction — sharing of genetic information between cells; it is facilitated by a virus, which carries DNA from one cell to another cell

Transformation — sharing of genetic information between cells as discrete fragments of DNA; a fragment from one cell is incorporated into the chromosome of another cell

Trophozoite — active, motile, feeding stage of a protozoa

Ultraviolet light — radiation having a wavelength shorter than that of visible light

Vaccine — suspension of weakened or dead microorganisms administered to prevent or treat an infectious disease

Vaccination — introducing all or part of a microorganism into an individual so that the antigens will stimulate production of antibodies and thereby confer immunity

Vector — any agent — usually insects (such as flies, lice, and biting bugs) or other arthropods (such as ticks and mites) — that carries disease from one host to another

Virology — study of viruses

Wavelength — measure of electromagnetic radiation and its energy level

White light — electromagnetic radiation in the visible range, such as sunlight or the light from an ordinary light bulb

Appendix B: Efficacy of Antibiotics Against Susceptible Pathogens

The following charts provide a ready reference to in vivo activity when the infecting organism has been identified. Empiric therapy often demands a powerful broad-spectrum antibiotic, in contrast to therapy designed to attack a specific invader. However, antimicrobial susceptibility testing should be routinely performed to ensure that an organism is indeed susceptible to the selected antibiotic.

KEY

1 Drug of choice as determined by comparative clinical efficacy or experience, susceptibility patterns, toxicity, cost, and practice patterns. In CNS infections, check package insert.

2 Alternative drug as determined by clinical practice, drug allergy, toxicity, and cost.

3 Drug with a low level of activity against this organism; variable or limited efficacy.

U Effective for treating urinary tract infections.

G Effective for GI infections.

☐ (Blank). Not indicated for this organism, or no information available for clinical evaluation.

NOTES TO CHARTS

A Cephalosporins, imipenem, and beta-lactamase inhibition combinations are not effective.

B Use penicillin or ampicillin plus an aminoglycoside for serious systemic infections due to enterococci.

C Antitoxin is primary therapy; antimicrobials are used only to decrease toxin production and to prevent a carrier state.

D Use ampicillin or penicillin with or without gentamicin. Preferred for bactericidal action.

E Spectinomycin is active against penicillin-resistant strains; however, resistance to this agent has been reported.

F Citrobacter freundii and C. diversus often have significantly different antibiotic sensitivity patterns. Speciation and susceptibility testing are particularly important.

G Use a tetracycline or chloramphenicol with or without streptomycin.

H For Haemophilus influenzae resistant to ampicillin, possible choices include chloramphenicol, a second- or third-generation cephalosporin, or trimethoprim-sulfamethoxazole, or a beta-lactamase-inhibitor combination agent.

I Combination therapy with an aminoglycoside and an antipseudomonal penicillin is warranted for serious systemic infections. Monotherapy with ceftazidime, cefoperazone, imipenem, or ciprofloxacin may be adequate in selected clinical settings.

J Vancomycin is effective orally only. Metronidazole is effective both orally and parenterally.

K Bacteroides bivius, B. distasonis, B. melaninogenicus, B. ovatus, B. thetaiotaomicron, B. uniformis.

L Isoniazid and rifampin are drugs of choice. However, nontubercular mycobacterial infections may require treatment with as many as five antibiotics.

M Contraindicated in children under age 18.

N The quinolones are currently not indicated for the treatment of meningitis.

O Norfloxacin is currently FDA-approved only for the treatment of urinary tract infections; however, clinical use includes treatment of susceptible prostatitis, gonorrhea, and infectious diarrhea and prophylactic bowel decontamination during neutropenia.

P Drug often used as co-drug to maximize bactericidal action or reduce resistance. Also indicated for prophylaxis or eradication of selected carrier states.

Q Tetracycline hydrochloride is preferred for most indications. Doxycycline is recommended for renal failure patients with infections outside the urinary tract for which a tetracycline is indicated. Tetracyclines are not recommended for pregnant women, infants, or children under age 8. A tetracycline class disk may be used to test susceptibility. However, minocycline may be effective against some strains of bacteria (for example, Acinetobacter, Enterobacteriaceae, and Staphylococcus aureus) resistant to other tetracyclines. No currently available tetracycline should be used in meningitis. Minocycline, rifampin, or ciprofloxacin is effective in the treatment of asymptomatic carriers of Neisseria meningitidis to eliminate meningococci from the nasopharynx.

Adapted with permission from Pharmacy Practice News, September 1991.

Appendix B: Efficacy of Antibiotics Against Susceptible Pathogens (continued)

AMINOGLYCOSIDES AND OTHER ANTIBIOTICS

Column groupings: **GRAM-POSITIVE AEROBES** (Cocci: columns 1–12; Bacilli: columns 13–16) and **GRAM-NEGATIVE AEROBES** (Cocci: columns 17–19; Enteric bacilli: columns 20–32).

Antibiotic	S. aureus (non-penicillinase)	S. aureus (penicillinase-producing)	S. aureus (methicillin-resistant[A])	S. epidermidis (non-penicillinase)	S. epidermidis (penicillinase-prod.)	S. epidermidis (methicillin-resistant[A])	Strep. group A (Strep. pyogenes)	Strep. group B	Strep. group D (enterococcal[B])	Strep. group D (nonenterococcal)	Strep. pneumoniae	Strep. viridans	Bacillus anthracis	Corynebacterium diphtheriae[C]	Corynebacterium jeikeium	Listeria monocytogenes[D]	Moraxella catarrhalis (Branhamella)	Neisseria gonorrhoeae[E]	Neisseria meningitidis	Citrobacter sp.[F]	Enterobacter sp.	Escherichia coli	Klebsiella pneumoniae	Morganella morganii	Proteus mirabilis	Proteus vulgaris	Providencia stuartii	Salmonella sp.	Salmonella typhi	Serratia sp.	Shigella sp.	Yersinia enterocolitica
AMINOGLYCOSIDES																																
AMIKACIN									3	3								2	2	1	1	2	2	2	2	2	1			1		2
GENTAMICIN	3	3	2	3	2	2			1	3								2	2	1	1	2	2	2	2	2	2	3	3	1	3	2
NETILMICIN																		2	2	1	1	2	2	2	2	2	2	3	3	1	3	2
STREPTOMYCIN									2	3																						
TOBRAMYCIN	3	3		3						3								2	2	1	1	2	2	2	2	2	2	3	3	2		2
OTHER ANTIBIOTICS																																
CHLORAMPHENICOL											2					2			2									2	1		2	
CLINDAMYCIN	2	2		2	2		2	2		3	2	3		3																		
ERYTHROMYCIN	2	2					2	2	3		2		2	1	2	2	2	3	3													
METRONIDAZOLE[M]																																
RIFAMPIN[P]	2	2	2	2	2	2									3				3													
SULFONAMIDES																			3	U	U	U	U	U	U				2		2	
TETRACYCLINES[M,Q]	3	3						2		2	2	2		2	2	2	2	3		3		3	3	3	2	2					2	2
TRIMETHOPRIM-SULFAMETHOXAZOLE	2	2	2	2	2	2												1	1	2	2	2	2	2	1	2	1	1	2	2	1	2
VANCOMYCIN	2	2	1	2	2	1	2	2	2	2	2	2		1																		
UTI AGENTS																																
CARBENICILLIN INDANYL	U						U	U	U											U	U	U		U	U	U	U			U		
NITROFURANTOIN	U						U	U	U											U	U	U	U									

Column groups: **GRAM-NEGATIVE AEROBES** — *Other bacilli* (Acinetobacter sp. … Yersinia pestis); **ANAEROBES** — *Gram-positive* (Clostridium difficile … Peptostreptococcus sp.) and *Gram-negative* (Bacteroides fragilis … Fusobacterium sp.); **Acid-fast**; **Actinomycetes**; **Chlamydia**; **Mycoplasma**; **Rickettsia**; **Spirochetes**.

Acinetobacter sp.	Aeromonas hydrophila	Bordetella pertussis	Brucella sp.[G]	Campylobacter jejuni	Francisella tularensis	Gardnerella vaginalis	Haemophilus ducreyi	Haemophilus influenzae[H]	Legionella pneumophila	Pasteurella multocida	Pseudomonas aeruginosa[I]	Pseudomonas cepacia	Pseudomonas mallei[G]	Pseudomonas maltophilia	Streptobacillus moniliformis	Vibrio cholerae	Yersinia pestis	Clostridium difficile[J]	Clostridium perfringens	Clostridium tetani	Peptococcus sp.	Peptostreptococcus sp.	Bacteroides fragilis	Bacteroides sp.[K]	Fusobacterium sp.	M. avium-intracellulare[L]	M. tuberculosis[L]	Actinomyces israelii	Nocardia sp.	Chlamydia psittaci	Chlamydia trachomatis	Mycoplasma pneumoniae	Ureaplasma urealyticum	Rickettsia sp.	Borrelia burgdorferi	Borrelia recurrentis	Leptospira sp.	Treponema pallidum
1	1			2	1		2			1	2	2	2			2										1	3		2									
1	1		2	2	1		2			1	2	2	2			2											3		2									
2	2			2	1		2			1	2	2	2			2													2									
					1			1									1									2	2											
1	2			2	1		2			1	2	2	2			2													2									
	2	2	2	2			2		2		2		2			2		2	2	2	2	2	2	2			3		2					2	2			
									2									2	2	2	2	2	2	2			3											
		1		1		1			1										3	3	3	3				1	3	2		1	1	1	2		1			3
									1																	1	2	2	3	3	1	1	1					
				1																						3	2											
				1							2					2																	1		1			3
2	3	1	2	2			2		2	2			2		1	2		3	2	2	2	2	3	2				2	2	1	1	1	2	1	1	1	2	2
1	2	2				1	2				1			1		2													1									
																		1																				
U																U																						

(continued)

Appendix B: Efficacy of Antibiotics Against Susceptible Pathogens (continued)

CEPHALOSPORINS AND QUINOLONES

Key
(*) first generation
(†) second generation
(**) third generation

Column groups: **GRAM-POSITIVE AEROBES** (Cocci C1–C12; Bacilli C13–C16) and **GRAM-NEGATIVE AEROBES** (Cocci C17–C19; Enteric bacilli C20–C32).

C1 = S. aureus (non-penicillinase); C2 = S. aureus (penicillinase-producing); C3 = S. aureus (methicillin-resistant[A]); C4 = S. epidermidis (non-penicillinase); C5 = S. epidermidis (penicillinase-prod.); C6 = S. epidermidis (methicillin-resistant[A]); C7 = Strep. group A (Strep. pyogenes); C8 = Strep. group B; C9 = Strep. group D (enterococcal[B]); C10 = Strep. group D (nonenterococcal); C11 = Strep. pneumoniae; C12 = Strep. viridans; C13 = Bacillus anthracis; C14 = Corynebacterium diphtheriae[C]; C15 = Corynebacterium jeikeium; C16 = Listeria monocytogenes[D]; C17 = Moraxella catarrhalis (Branhamella); C18 = Neisseria gonorrhoeae[E]; C19 = Neisseria meningitidis; C20 = Citrobacter sp.[F]; C21 = Enterobacter sp.; C22 = Escherichia coli; C23 = Klebsiella pneumoniae; C24 = Morganella morganii; C25 = Proteus mirabilis; C26 = Proteus vulgaris; C27 = Providencia stuartii; C28 = Salmonella sp.; C29 = Salmonella typhi; C30 = Serratia sp.; C31 = Shigella sp.; C32 = Yersinia enterocolitica

CEPHALOSPORINS

Drug	C1	C2	C3	C4	C5	C6	C7	C8	C9	C10	C11	C12	C13	C14	C15	C16	C17	C18	C19	C20	C21	C22	C23	C24	C25	C26	C27	C28	C29	C30	C31	C32
CEFADROXIL (*)	1	1		1	1		2	2		2	2	2										1	1		1							
CEFAZOLIN (*)	1	1		1	1		2	2		2	2	2										1	1		1							
CEPHALEXIN (*)	1	1		1	1		2	2		2	2	2										1	1		1							
CEPHALOTHIN (*)	1	1		1	1		2	2		2	2	2.										1	1		1							
CEPHAPIRIN (*)	1	1		1	1		2	2		2	2	2										1	1		1							
CEPHRADINE (*)	1	1		1	1		2	2		2	2	2										1	1		2							
CEFACLOR (†)	2	2		2	2		2	2		2	2	2					2	2				1	1		2							
CEFAMANDOLE (†)	2	2		2	2		2	2		2	2	2					2	2				1	1		2							
CEFMETAZOLE (†)	2	2		3	3		2	2		2	2	2					2	1				1	1	2	2	2	1				2	2
CEFONICID (†)	3	3		3	3		2	2		2	2	2					2	2				1	1		2							
CEFOTETAN (†)	3	3		3	3		2	2		2	2	2					2	1				1	1	2	2	2	1			1		2
CEFOXITIN (†)	2	2		3	3		2	2		2	2	2					2	1				1	1	3	2	2	1				3	
CEFUROXIME (†)	2	2		2	2		2	2		2	2	2					2	1	2			1	1		2							
CEFUROXIME AXETIL (†)	2	2		2	2		2	2			2	2					2	2				1	1		2							
CEFIXIME (**)							2	2			2	2					1	1			2	1	1	3	2	1	1				2	
CEFOPERAZONE (**)	2	2		3	3		2	2		2	2	2					2	1		1	2	1	1	1	2	1	1			2	1	
CEFOTAXIME (**)	2	2		2	2		2	2		2	2	2					1	1	2	1	2	1	1	1	2	1	1			1		2
CEFTAZIDIME (**)	3	3		3	3		2	3		2	3	2					2	1		1	2	1	1	2	1	1				1		
CEFTIZOXIME (**)	2	2		2	2		2	2		2	2	2					1	1		1	2	1	1	1	2	1	1			1		2
CEFTRIAXONE (**)	2	2		2	2		2	2		2	2	3					1	1	2	1	2	1	1	1	2	1	1		2	1		2

QUINOLONES

Drug	C1	C2	C3	C4	C5	C6	C7	C8	C9	C10	C11	C12	C13	C14	C15	C16	C17	C18	C19	C20	C21	C22	C23	C24	C25	C26	C27	C28	C29	C30	C31	C32
CIPROFLOXACIN[N,O]	2	2	2	2	2	2	3	3	3	3	3	3				2	2	2	2	1	1	1	1	1	1	1	1	1	2	1	1	2
NORFLOXACIN[N,O,P]	U	U	U	U	U	U		U	U	U						U		U	U	G	U	U	U	U	U	U	G		U	G	G	
OFLOXACIN[N,O]	2	2	2	2	2	2	3	3	3	3	3	3				2	2	2	2	1	1	1	1	1	1	1	1	1	2	1	1	2

	GRAM-NEGATIVE AEROBES																		ANAEROBES								Acid-fast		Actinomycetes		Chlamydia		Mycoplasma		Rickettsia	Spirochetes			
	Other bacilli																		Gram-positive					Gram-negative															
	Acinetobacter sp.	Aeromonas hydrophila	Bordetella pertussis	Brucella sp.[G]	Campylobacter jejuni	Francisella tularensis	Gardnerella vaginalis	Haemophilus ducreyi	Haemophilus influenzae[H]	Legionella pneumophila	Pasteurella multocida	Pseudomonas aeruginosa[I]	Pseudomonas cepacia	Pseudomonas mallei[G]	Pseudomonas maltophilia	Streptobacillus moniliformis	Vibrio cholerae	Yersinia pestis	Clostridium difficile[J]	Clostridium perfringens	Clostridium tetani	Peptococcus sp.	Peptostreptococcus sp.	Bacteroides fragilis	Bacteroides sp.[K]	Fusobacterium sp.	M. avium-intracellulare[L]	M. tuberculosis[L]	Actinomyces israelii	Nocardia sp.	Chlamydia psittaci[M]	Chlamydia trachomatis[M]	Mycoplasma pneumoniae[M]	Ureaplasma urealyticum[M]	Rickettsia sp.	Borrelia burgdorferi	Borrelia recurrentis	Leptospira sp.	Treponema pallidum
									2																														
									2		2										2	2	2																
									2																														
									2		2																												
									2		2																												
									2																														
									2																														
									2		2																												
		2							2		2									2	2	2	2	2	2	2													
									2																														
		2							2		2									2	2	2	2	2	3	2													
		3							2		2									2	2	2	2	2	2	2													
		2							1											3	3	2	2																
									2		2									2	2	2	2																
									1																														
	3	2							1		2	2	2		2					2	2	2	2	3	3	3													
	3	1						2	1		1	3								2	2	2	2	3	3	3				2						1			
	2	2							1			1	2		2					3	3	3	3																
	3	1							1		2	U								2	2	2	2	3	3	3													
	3	1						1	1		2	3																								1			3
	2	1		1	1		2	1	1	2	2	2	3		2		2										3	3				2	3	3	3				
		G			G							U	U																					U					
	2	1		1	1		2	1	1	2	2	2	3		2		2										3	3				2	3	3	3				

(continued)

Appendix B: Efficacy of Antibiotics Against Susceptible Pathogens *(continued)*

PENICILLINS AND OTHER BETA-LACTAM ANTIBIOTICS

Column key (GRAM-POSITIVE AEROBES — Cocci, Bacilli; GRAM-NEGATIVE AEROBES — Cocci, Enteric bacilli):

1. S. aureus (non-penicillinase)
2. S. aureus (penicillinase-producing)
3. S. aureus (methicillin-resistant)[A]
4. S. epidermidis (non-penicillinase)
5. S. epidermidis (penicillinase-prod.)
6. S. epidermidis (methicillin-resistant)[A]
7. Strep. group A (Strep. pyogenes)
8. Strep. group B
9. Strep. group D (enterococcal)[B]
10. Strep. group D (nonenterococcal)
11. Strep. pneumoniae
12. Strep. viridans
13. Bacillus anthracis
14. Corynebacterium diphtheriae[C]
15. Corynebacterium jeikeium
16. Listeria monocytogenes[D]
17. Moraxella catarrhalis (Branhamella)
18. Neisseria gonorrhoeae[E]
19. Neisseria meningitidis
20. Citrobacter sp.[F]
21. Enterobacter sp.
22. Escherichia coli
23. Klebsiella pneumoniae
24. Morganella morganii
25. Proteus mirabilis
26. Proteus vulgaris
27. Providencia stuartii
28. Salmonella sp.
29. Salmonella typhi
30. Serratia sp.
31. Shigella sp.
32. Yersinia enterocolitica

β-LACTAMASE-SUSCEPTIBLE (Nonantipseudomonal)

Drug	1	2	3	4	5	6	7	8	9	10	11	12	13	14	15	16	17	18	19	20	21	22	23	24	25	26	27	28	29	30	31	32
AMOXICILLIN	2			2			2	1		2	2	2					1	2				2			1			1	2			
AMPICILLIN	2			2			2	1	1	2	2	2				1	1	2				2			1			1	2			2
BACAMPICILLIN	2			2			2	1		2	2	2					3					2			1			1	2			
PENICILLIN G	1			1			1	1	1	1	1	1	1	2		1	1	1														
PENICILLIN V	1			1			1	1	2	1	1	1	1	2			3															

β-LACTAMASE-SUSCEPTIBLE (Antipseudomonal)

Drug	1	2	3	4	5	6	7	8	9	10	11	12	13	14	15	16	17	18	19	20	21	22	23	24	25	26	27	28	29	30	31	32
MEZLOCILLIN	3			3			2	1	2	3	2	3						2		2	2	2	2	2	2	2	2			2		
PIPERACILLIN	3			3			2	1	2	3	2	3						2		2	2	2	2	2	2	2	2			2		
TICARCILLIN	3			3			2	2		3	2	3						2			2			2	2	2	2			2		

β-LACTAMASE-RESISTANT (Antistaphylococcal)

Drug	1	2	3	4	5	6	7	8	9	10	11	12	13	14	15	16	17	18	19	20	21	22	23	24	25	26	27	28	29	30	31	32
CLOXACILLIN	2	1		2	1		3					3																				
DICLOXACILLIN	2	1		2	1		3					3																				
METHICILLIN	2	1		2	1		3					3																				
NAFCILLIN	2	1		2	1		3					3																				
OXACILLIN	2	1		2	1		3					3																				

β-LACTAMASE-RESISTANT (Others)

Drug	1	2	3	4	5	6	7	8	9	10	11	12	13	14	15	16	17	18	19	20	21	22	23	24	25	26	27	28	29	30	31	32
AMOXICILLIN/CLAVULANATE	2	2	3	2	2	3	2	1	3	2	2	2				2	1	1	2			2	2		2			1	2			
AMPICILLIN/SULBACTAM	2	2	3	2	2	3	2	2	2	2	2	2				2	1	1	2			2	2	2	2			1	2			2
TICARCILLIN/CLAVULANATE	2	2		3	2		2	2		3	2	3					2	2	2		2	2	2	2	2	2	2			2		
AZTREONAM																		2		2	2	2	2	2	2	2	2			2		
IMIPENEM/CILASTATIN	2	2	3	2	2	3	2		2		2						2	2	3	1	1	2	2	1	2	1	1			2		

| | GRAM-NEGATIVE AEROBES | | | | | | | | | | | | | | | | | | ANAEROBES | | | | | | | | Acid-fast | | Actinomycetes | | Chlamydia | | Mycoplasma | | Rickettsia | Spirochetes | | | |
											Other bacilli								Gram-positive					Gram-negative															
	Acinetobacter sp.	Aeromonas hydrophila	Bordetella pertussis	Brucella sp.[G]	Campylobacter jejuni	Francisella tularensis	Gardnerella vaginalis	Haemophilus ducreyi	Haemophilus influenzae[H]	Legionella pneumophila	Pasteurella multocida	Pseudomonas aeruginosa[I]	Pseudomonas cepacia	Pseudomonas mallei[G]	Pseudomonas maltophilia	Streptobacillus moniliformis	Vibrio cholerae	Yersinia pestis	Clostridium difficile[J]	Clostridium perfringens	Clostridium tetani	Peptococcus sp.	Peptostreptococcus sp.	Bacteroides fragilis	Bacteroides sp.[K]	Fusobacterium sp.	M. avium-intracellulare[L]	M. tuberculosis[L]	Actinomyces israelii	Nocardia sp.	Chlamydia psittaci	Chlamydia trachomatis	Mycoplasma pneumoniae	Ureaplasma urealyticum	Rickettsia sp.	Borrelia burgdorferi	Borrelia recurrentis	Leptospira sp.	Treponema pallidum
			2		2	1																2	2																	
			2	2	2	1																2	2								1	2							
			2		2	1																2	2																
												1						1	1	1	1	1				1		1								1	2	1	1
												1						1	1	1	1	1				1		1								1	2	1	1
	2											3	2		1				2	2	2	2	2	3															
	2											3			1				2	2	2	2	2	3															
	2											3	2		1				2	2	2	2	2	3	2														
		2										2	2	1	2		2		2	2	2	2																	
	3											1							2	2	2	2	1	1	1														
	2											2	2		2		2		2	2	2	2	1	1	2														
	3	2										2					2																							
	1	2	2									2					2		2	2	2	2	1	1	1				2										

Selected References

Cundy, K.R., et al. *Infection Control: Dilemmas and Practical Solutions.* New York: Plenum Publishing Co., 1990.

Finegold, S.M., and Baron, E.J. *Bailey and Scott's Diagnostic Microbiology,* 8th ed. St. Louis: C.V. Mosby Co., 1990.

Joklik, W.K., et al. *Zinsser Microbiology,* 19th ed. New York: Appleton and Lange, 1988.

Leventhal, R., and Cheadle, R.F. *Medical Parasitology,* 3rd ed. Philadelphia: F.A. Davis Co., 1989.

Mulvihill, M.L. *Human Diseases: A Systemic Approach,* 3rd ed. Norwalk, Conn.: Appleton and Lange, 1990.

Pelczar, M.J., and Chan, E.C. *Microbiology,* 5th ed. New York: McGrawHill Book Co., 1985.

Sherris, J.C., et al. *Medical Microbiology: An Introduction to Infectious Diseases,* 2nd ed. New York: Elsevier Science Publishing Co., 1990.

Volk, W.A., and Wheeler, M. *Basic Microbiology,* 6th ed. New York: Harper Collins, 1987.

Index

A

Acid-fast stain, 11, 20-21
Actinomadura, 71
Adenosine triphosphate, 24, 25
Adenovirus, 57-58
Aerobes, 19, 25, 110t-115t
Agar, 17
Algae, 81, 82i
Alternaria, 68-69
Amebae, 72, 73, 74i, 75
Aminoglycosides, 103, 110t-111t
Amoeba, 74
Anabolism, 23
Anaerobes, 19, 25, 110t-115t
Antibiotics, 5, 103, 109t-115t
Antibody, 35, 94
Antigen, 35, 94
Antimicrobial agents, 34
Antiseptics, 4, 101
Arenavirus, 59
Arthrospores, 66
Ascaris, 86
Aschelminthes, 83, 84
Ascospores, 66
Aspergillus, 69
ATP. *See* Adenosine
 triphosphate.
Autoclave, 99
Autotrophs, 24

B

Bacilli, 14, 15i
 gram-negative, 45, 46t,
  47-48, 110t-115t
 gram-positive, 41, 42t, 43-44,
  110t-115t
 gram-variable, 48-49
Bacillus, 44
Bacteria, 36-50
 atmospheric requirements of,
  19t
 cell structure of, 15-16
 definition of, 1
 morphology of, 14, 15i
 sources of, 19t
Bacteriophage, 58
Bacteroidaceae, 45
Balantidium, 78
Bdellovibrio, 50
Bivirus, 60
Blastomyces, 70
Blastospores, 66
Bordetella, 47
Broth, 17, 106
Brucella, 47
Budding, 16, 28
Bunyavirus, 59

C

Candida, 67
Capnophiles, 19
Capsid, 55
Capsule, 15
Capsule stain, 11, 21
Catabolism, 23-24
Catalase, 30

Cell
 as defense against infection,
  94
 definition of, 1
 eukaryotic, 16
 prokaryotic, 16
Cell theory, 3
Cephalosporins, 103, 112t-113t
Cestoda. *See* Tapeworms.
Chemicals, infection control
 and, 100
Chemosynthesis, 24
Chemotaxis, 74
Chemotherapy, 4
Chlamydia, 53, 110t-115t
Chlamydiaceae, 54
Chlamydomonas, 82
Chlamydospores, 66
Chlorophyll, 76
Chlorophyta. *See* Algae.
Chromatin, 73
Chromosomes, 15, 28
Cilia, 33, 72, 77
Ciliates (Ciliata), 72, 77, 78i, 79
Clonorchis, 90
Clostridium, 44
Cocci, 14, 15i
 gram-negative, 39, 40t, 41,
  110t-115t
 gram-positive, 37, 38t, 39,
  110t-115t
Coccidioides, 70
Coenzymes, 29, 30
Commensalism, 2
Conidiophores, 66
Conjugation, 28
Copepods, 88
Coronavirus, 59-60
Corynebacteriaceae, 43
Cotton blue stain, 66
Coxsackieviruses, 60
Cryophiles, 17
Cryptococcus, 67
Cultures, 14, 16-17, 18t, 19,
 33-34
Cyst, 73
Cytomegalovirus, 58
Cytoplasm, 16

D

Darkfield microscopy, 6
Deaminase, 30
Decarboxylase, 30
Dehydrogenase, 30
Denaturation, 99
Deoxyribonucleic acid, 28, 35
 viruses associated with,
  57-59
Desiccation, 100
Differential media, 17, 18t
Diffusion, 79
Dinoflagellata, 75-76
Diphyllobothrium, 88
Disease transmission, 97
Disinfectants, 98, 101
DNA. *See* Deoxyribonucleic
 acid.

E

Echoviruses, 60
Ehrlich, Paul, 4
Elementary bodies, 54
Embden-Meyerhof pathway, 25,
 27i
Endotoxins, 96
Enriched media, 17, 18t
Enrichment media, 17, 18t
Entamoeba, 74
Enterobacteriaceae, 47
Enterobius, 86
Enteroviruses, 60
Enzymes, 29-30
Epidemiology, 96-97
Epstein-Barr virus, 58
Escherichia coli
 identification of, 35
 taxonomy of, 32t
Euglena, 76
Exotoxins, 96

F

Facultative organisms, 19, 25
Fasciola, 91
Fermentation, 25, 106
Filaria, 85, 86
Fission, 16, 28
Flagella, 15, 72
Flagella stain, 21
Flagellates, 72, 75, 76i, 77
Fleming, Alexander, 5, 103
Flora, 36, 94
Flukes, 89, 90i, 91
Fluorescent stain, 21
Fomites, 97
Foraminifera, 74
Fracastoro, Girolamo, 3
Francisella, 47-48
Fungi, 1
 dimorphic, 69-70
 monomorphic, 66-69
 morphology of, 64, 65i, 66
 reproduction of, 66
 types of, 64
 of uncertain designation,
  70-71

G

Gametes, 28
Gametocytes, 28, 73
Gardnerella, 49
Genes, 28
Genetics, 28
 alterations in, 29
Genus, 31, 32-33
Geotrichum, 67
Germ theory, 4
Giardia, 76
Giemsa stain, 21
Glycolysis, 25
Gomori methanamine silver
 stain, 21
Gram stain, 11, 20, 37

H

Haemophilus, 48

i refers to an illustration; t, to a table.

Haemophilus aegyptius, 33
Halogens, 102
Helminths, 1, 2, 83-91
Hemin, 48
Hemolysis, 34, 37
Hepatitis B virus, 58
Herpesvirus, 58
Heterotrophs, 24
Histoplasma, 70
Hooke, Robert, 3
Hookworms, 86-87
Host, 2, 16, 56-57
Humans, taxonomy of, 32t
Hydrolase, 30
Hymenolepis, 89
Hyphae, 65i

I

Immunity
    active, 102
    passive, 103
Incineration, 100
Inclusion bodies, 57
Incubation, 57
Infection
    defenses against, 93-94
    nosocomial, 96
    prevention of, 97, 102
    sources of, 96
    treatment of, 103
Infectivity, 94-95
Influenzavirus, 60
Inoculum, 19
Intermediate organisms, 2,
    51-54

J

Janssen, Johannes and
    Zacharias, 3
Jenner, Edward, 4
Jobbt, Louis, 4

K

Kinase, 30
Koch, Robert, 4, 95
Koch's Postulates, 95
Krebs cycle, 25, 26i

L

Lactobacillus, 44
Larvae, 83
Leishmania, 76
L-forms, 51-53
Linnaeus, Carolus, 31
Lipase, 30
Lister, Joseph, 4
Listeria, 44
Listeria monocytogenes, 33
Lyophilization, 100
Lysis, 58

M

Macronucleus, 77
Maduromycetes, 68
Mastigophora. *See* Flagellates.
Media, culture, 16-17, 18t
Mesophiles, 17
Messenger RNA, 28
Metabolism, 23-24

Microaerophiles, 19
Microbiology, history of, 3-5
Micrococcaceae, 39
Micronucleus, 77
Microorganisms
    control of, 98-103
    culturing of, 16-19
    definition of, 1
    growth of, 22-30
    identification of, 31-35
    morphology of, 11
    motility of, 11
    pathogenic nature of, 95
    reproduction of, 26, 28-29
    respiration in, 25
    study of, 2-3
    types of, 1-2
Microscope, 14
    acoustic, 7-8
    compound, 3, 7
        parts of, 8, 9i, 10-11
    electron, 5, 7
    fluorescence, 7
    light, 6-7
        use of, 12
    magnification with, 8
    resolution with, 8
    ultraviolet, 7
Microsporum, 69
Moist heat, 99
Molds, 2, 64, 68-69
Mosaic viruses, 60
Mutations, genetic, 29
Mutualism, 2
Mycelium, 65i
Mycobacteriaceae, 43
Mycology, 3, 64
Mycoplasmataceae, 53,
    110t-115t

N

Naegleria, 74
Needham, John, 4
Negative stain, 20
Neisseriaceae, 41
Neisseria gonorrhoeae, 41
Neisseria meningitidis, 41
Nematoda. *See* Roundworms.
Nitrobacteraceae, 47
Nitrogen fixation, 45
Nitrosomonas, 33
Nocardiaceae, 43
Nomenclature, 31-32
Norwalk viruses, 62
Nucleotides, 28
Nucleus, 1
Nutrient media, 17

O

Obligate organisms, 19, 53-54
Opportunists, 2, 96
Organelles, 1, 16
Organisms
    facultative, 19, 25
    intermediate, 2, 51-54
    obligate, 19, 53-54
Osmotic pressure, 100
Oxidase, 30
Oxidation, 25

P

Papovavirus, 58
Paragonimus, 91
Paramecium, 79
Paramyxovirus, 60
Parasites, 2
    obligate intracellular, 53-54
Parasitism, 2
Parasitology, 3
Parthenogenesis, 83
Pasteur, Louis, 4
Pasteurella, 48
Pasteurization, 4, 99
Pathogen, 2
Penicillin, 5, 103, 114t-115t
Penicillium, 69
Peptococcaceae, 39
Periodic acid-Schiff stain, 66
Phagocytosis, 74
Phase-contrast microscopy, 6
Photosynthesis, 24
Phycomycetes, 68
Picornavirus, 60
Pili, 15
Pinworms, 85, 86
Planaria, 91
Plasmodium, 79, 80
Pneumocystis, 80-81
Poliovirus, 60
Positive stain, 20
Poxvirus, 58
Prions, 62
Proglottids, 88
Propionibacteriaceae, 43
Propionibacterium acnes, 33
Protein infectious particles, 62
Protist, 72
Protozoa, 1, 2, 72-81
Pseudomonadaceae, 47
Pseudopodia, 72
Psychrophiles, 17

Q, R

Quinolones, 112t-113t
Radiation, 100
Receptor sites, 56
Recombination, genetic, 29
Reed, Walter, 4
Reovirus, 60
Replication, 28
Reproduction, microorganisms
    and, 26, 28-29, 66
Respiration, microorganisms
    and, 25
Reticulate bodies, 54
Retrovirus, 60-61
Rhabdovirus, 61
Rhinoviruses, 60
Rhizopoda, 72-75
Ribonucleic acid, 28
    viruses associated with,
        59-61
Ribosome, 28
Ricketts, Howard Taylor, 54
Rickettsia, 53, 110t-115t
Rickettsiaceae, 54
RNA, 28, 59-61
Rotavirus, 60
Rotifer, 84

i refers to an illustration; t, to a table.

Roundworms, 84, 85i, 86-88
Ruska, Ernst, 5

S

Saccharomyces, 67
Salk, Jonas, 5
Sanitation, 100
Saprophytes, 24
Schistosoma, 91
Schizogony, 80
Scolex, 88, 89i
Selective media, 17, 18t
Sepsis, 39
Smallpox, 4
Smears, 11
Sonication, 100
Spallanzani, Lazzaro, 4
Species, 31, 32-33
Spirillaceae, 49
Spirochaetaceae, 50
Spirochetes, 14, 49-50,
    110t-115t
Spontaneous generation, 4
Sporangia, 66
Sporangiospores, 66
Spores, 21, 44, 66
Spore stain, 21
Sporogony, 80
Sporothrix, 70
Sporozoans (Sporozoa), 72, 79,
    80i, 81
Stains, 7, 11, 14
    techniques for, 20-21
Staphylococcus aureus, 33, 37,
    39
Staphylococcus hominis, 39
Sterigmata, 69
Sterilization, 98, 99
Streptobacillus, 48
Streptococcaceae, 39
Streptococcus malodoratus, 33
Streptococcus pneumoniae, 39
Streptococcus pyogenes, 34t,
    37, 39
Streptococcus viridans, 39
Streptomycetaceae, 44
Strongyloides, 87
Substrate, 29
Symbiosis, 2
Synergy, 103
Syphilis, 4

T

Taenia, 89
Tapeworms, 88, 89i
Taxonomy, 31, 32t
Tetracyclines, 103
Thallospores, 66
Thermophiles, 17
Togavirus, 61
Toxocara, 85, 87
Toxoplasma, 81
Transcription, 28
Transduction, genetic, 29
Transformation, genetic, 29
Translation, 29
Transport media, 17
Trematoda. *See* Flukes.
Trichinella, 85, 87

Trichomonas, 77
Trichonympha, 77
Trichosporon, 67-68
Trichuris, 87
Trophozoite, 73
Trypanosoma, 77
Tuberculosis, 4
Turbatrix, 87-88

U, V

Ultraviolet light, 100
Vaccination, 4
Vaccine, 4
van Leeuwenhoek, Antonie, 4
Varicella-zoster virus, 58
Vectors, 54, 97
Veillonellaceae, 41
Vibrionaceae, 47
Virion, 56i
Virology, 3, 55-57
Virulence, 95-96
Viruses, 1, 2, 55-62
    DNA, 55-59
    hosts of, 56-57
    morphology of, 55, 56i
    prions and, 55, 62
    RNA, 55-56, 59-61
    of uncertain designation,
        61-62
Volvox, 81, 82

W, X, Y, Z

Wright stain, 21
Yeasts, 2, 64, 66-68
Yellow fever, 4

i refers to an illustration; t, to a table.

Notes